GW01158607

Charles William Lynn CBE1941 (1908-1985)

The Long Garden Master in the Gold Coast;

Life and times of a Colonial Agricultural Officer
in the Gold Coast,
1929-1947

Charles, Marjorie and Sylvia Lynn

Authors online

An Authors OnLine Book

Copyright © Charles, Marjorie and Sylvia Lynn 2012

Cover image:
Department of Agriculture Staff at Tamale 1946 Dady (AAO), Akenhead
(Dagomba) C.W.Lynn (Chief AO of N.T's) Smith(Mamprusi) Hinds (Lawra)

All rights reserved. No part of this publication may be reproduced, stored in
a retrieval system, or transmitted in any form or by any means, electronic,
mechanical, photocopy, recording or otherwise, without prior written
permission of the copyright owner. Nor can it be circulated in any form
of binding or cover other than that in which it is published and without
similar condition including this condition being imposed on a subsequent
purchaser.

British Library Cataloguing Publication Data.
A catalogue record for this book is available from the British Library

ISBN 978-0-7552-0710-7

Authors OnLine Ltd
19 The Cinques
Gamlingay, Sandy
Bedfordshire SG19 3NU
England

This book is also available in e-book format, details of which are
available at www.authorsonline.co.uk

Dedicated to:

The men and women who Served as Colonial Officers in the Northern Territories of the Gold Coast, and to Professor Meyer Fortes and his wife Sonia who as resident Social Anthropologists, Tongo, enlightened them in their endeavours to bring about Indirect Rule.

"I always think that we live our lives, spiritually, by what others have given us in the significant hours of our life. These significant hours do not announce themselves as coming, but arrive unexpectedly. Nor do they make a great show of themselves; they pass almost unperceived. Often, indeed, their significance comes home to us first as we look back, just as the beauty of a piece of music or a landscape often strikes us first in our recollection of it. Much that has become our own in gentleness, modesty, kindness, willingness to forgive, in veracity, loyalty, resignation under suffering, we owe to the people in whom we have seen or experienced these virtues at work, sometimes in a great matter, sometimes in a small.

A thought that had become act sprang into us like a spark, and lit a new flame within us. If we had before us those who have thus been a blessing to us, and could tell them how it came about, they would be amazed to learn what passed over from their lives into ours."

Albert Schweitzer (1875-1965)

CONTENTS

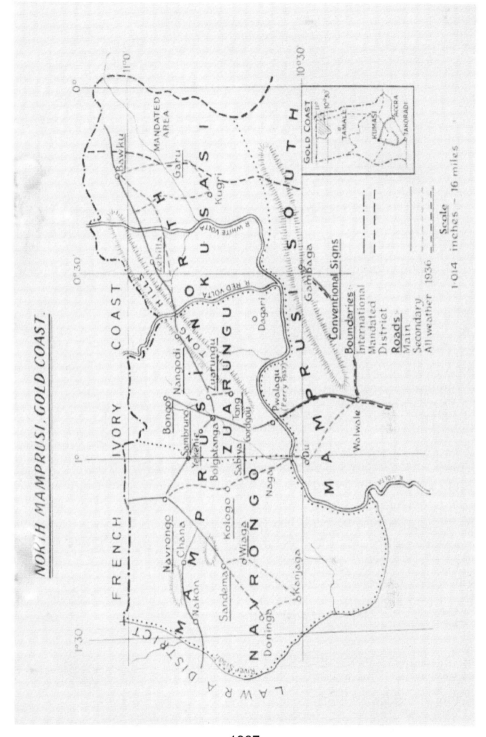

NORTH MAMPRUSI, GOLD COAST.

1937

1

PREFACE

In 1968, while redecorating my grandmother's flat, I came across an old battered brown suitcase containing the letters my father had written to her from the Gold Coast. I had lived in Tamale as a 3 year-old child, and our closest family friends were from that period of my father's life. So I grew up with the stories and affection for Tamale, my first home – and a place called Zuarungu. In later life I encouraged my father to write his memoirs, reminding him of half- remembered stories and encouraging him to dig deep into his memory. He died in 1985 leaving me to finish the task.

My somewhat unorthodox idea was to integrate the most interesting letters into the memoirs. After my parents' marriage in 1937 my mother too wrote letters home to my grandmother, and these also deserve a place in the book. So our family book "The Long Garden Master" came to be written. I knew many of the people mentioned in this book, and have inherited some of their personal letters addressed to my father, Charles.

To further illuminate the times I included background details, using references and quotes from the daily personal diaries of Professor Meyer Fortes and his wife Sonia. I use long quotes to get the authentic voices of the era. Fortes was doing his fieldwork on the Tallensi from January 1934 to April 1937. These excerpts are not from his fieldwork books, but give the more personal psychological details of the people and situations as he found them. As a trained psychologist his observances are very much of interest, as a reflection of those times.

Slavery was on the increase in the Sixteenth Century, especially in the Western Mediterranean from Tunis to Algiers, along the area known as the Barbary Coast. The Berbers had created a flourishing state along the edge of the Mediterranean Sea, and used slaves to build it. Some of these slaves were African, but many were white Christians raided from Europe or passing trading ships. In 1591 they also raided and ended the Songhai Empire on the Niger River. Portugal and Spain were the first Europeans to take slaves to the Caribbean and South America, especially for mining. The first British trading charters for the West Coast of Africa were given in 1618, and the slave trade soon began here too, driven by the need for labour in the new American Colonies. Slavery had never been a normal activity amongst the British before the Transatlantic trade. It soon became a very important and valuable trade as the demand for sugar and cotton grew. Before long great Atlantic ports were developed to handle the trade, especially Liverpool and Bristol. In 1807 the British ended its own trade by abolition. The Royal Navy, who had been its protector, now had dedicated anti slave trade squadrons. Considerable money and effort was put into catching and prosecuting slave traders. However it did continue despite their best efforts. There was no end to the supply in Africa, indeed as late as 1887, a British envoy saw between 500 and 2,000 slaves in Salaga Market. It was estimated even then that an average of 20,000 changed hands annually. With the rise in social conscience in the general public, especially from religious groups, came abolition . In the Gold Coast gold and Ivory were still important exports and a new cocoa industry was then established. In 1924, Roses began exporting lime juice back to England in wooden barrels to make their famous cordial. Sir Gordon Guggisberg, meeting Lauchlin Rose during the First World War, personally got investment and development for this valuable export crop.

The two Ashanti Wars at the end of the nineteenth century were costly affairs to the British, affecting the coastal and forest areas, well known to the Europeans conducting the slave trade. Charles had little to do with this area, except for a short time at Ejura in 1933, spending all his time in the Savannah area of the Protectorate, known as the Northern Territories.

This area was, before British rule, dominated by the slave raiders, Samory in the west and Babatu in the east. For nearly three decades these freebooters decimated the area's political and financial stability, until defeated by the Europeans around 1898. Though they were Islamic in tradition, this was not an invasion for "jihad", as had taken place in Northern Nigeria. They traded slaves for guns imported from the coastal ports, despite the end of the Atlantic slave trade. With the coming of British Rule many domestic slaves were released, and those that could returned home; waged labour from the north was to replace them. The remnants of these slave- raiding armies, many of Moshi descent from what was then Upper Volta, were amalgamated into the Mounted Force. They were then used in pacifying the North, especially the north eastern region, the Zuarungu District, the main area of my book. Some of these men were still alive in 1930 holding important positions,often as interpreters.

The early colonial officers laid the foundations of local government, firstly by military control and then by direct rule. They were well chosen, able men. One such man was Captain Cecil H. Armitage, a well respected officer who arrived in Ashanti in 1894, and who was Assistant Inspector of the Gold Coast Constabulary in Kumasi. He was involved in the Ashanti Expedition of 1895. Frequently he visited the North on military matters, as a travelling Commissioner, concerning the unrest in the Tong Hills, and became its Chief Commissioner in 1910. His father was doctor to Florence Nightingale, social reformer of hospital sanitary conditions and the daughter of an ardent anti-slavery father. Armitage was heard to boast that she regarded him as her adopted son. One of their discussions, no doubt, would have been with regard to the evils of transatlantic and domestic slavery, and the effects of Babatu and Samory.

The Boer War was also a factor that affected the Gold Coast when men, some of whom later worked with Charles, were called away to fight. Lieutenant Colonel Henry P. Northcott was Commissioner and Commandant of the Northern Territories in 1897, but died in South Africa in 1899. He was a very able man who wrote one of the earliest published colonial books on the north, "Report on the Northern Territories of the Gold Coast". With him in the North was Arthur H. C. Walker Leigh, who joined the British Constabulary in 1898. He served with Northcott that same year in the Fra Fra expedition, fighting in the Boer War briefly, before returning to take part in the Tallensi expedition in 1901. He was Chief Commissioner Northern Territories and retired only a few months before Charles arrived. Percy Whittall fought in the Boer War, joining the West African Frontier Force in 1907; he became Superintendent of Roads in Ashanti and the Northern Territories. Charles met him just before he retired, and gave a home to his cat which he imaginatively called Percy! There were many others who were well remembered, and had made important decisions that prevailed till the change to indirect rule.

A. E. Watherston in 1900 was the Chief British Commissioner Anglo-French Boundary Commission in the north, becoming District Commissioner in 1905. He gave a lecture to the African Society in London in June 1908, in which he described that while mapping the northern boundary in 1900 he found the bush around Bole "in many places full of skeletons", with a rather depleted population. Samory had endeavoured to maintain himself in the northwest by using domestic slaves to grow crops, but his power was weakened by famine and sickness. In the adjacent north east, a very populous area, Babatu confined himself to raiding. The poverty of the area meant there was no wealth to buy guns or form an army; instead the people of North Mamprusi waged pitch battles with the invaders using bows and poisoned arrows. Sometimes the people themselves would raid each other, going home at night; hence there was no unity to fight off raiders. Long distance traders had to travel well armed, and were often attacked. In 1907, Gambaga and Navoro (the spelling at the time), and Bawku in 1908, were made into "stations" with a small constabulary and mounted section. Messages took 16 days at the time to reach Accra from Tamale, and 16 days back. Watherston found the Chiefs incapable of making anybody obey them, as that was not the Chief's traditional role. His colleagues meanwhile questioned the remuneration that Britain would get out of the country, for all the trouble and money spent. Another asked what would be the effect on trade if all European powers abandoned their "unhealthy sub-tropical countries"? He reflected that people in England had high expectations of their instilling in the local people "civilising" and "intellectual development". "We are obliged to raise the native to a higher standard of civilisation in accordance with our European ideas." Watherston died suddenly of blackwater fever just over a year later in 1910, in Tamale. The North once again lost a good man.

Important people in the early days were surveyors. They got to know areas of the country in great detail, and interacted with the local people in ways different from those in the administration. They therefore proposed imaginative development projects to open up trade and advance the economy. Military Officer, Surveyor, and finally Governor, Brigadier-General Sir Frederic Guggisberg joined the Service in 1902. From 1905–8, he meticulously mapped out the Coast and Ashanti regions, as he records in his book "We Two in West Africa"; he went on to do the same in Nigeria. When he became Governor of the Gold Coast from 1919-27 one of his main aims was to form the groundwork for development of the whole country. He put forward a bold plan to build a circular loop for the railway line running north, perhaps linking the French Territory to the north and so on to Nigeria. It would clearly have been a very difficult job, but would have brought much needed development north as the roads at the time were very poor. It would also have stimulated the search for agricultural produce from the north. One of the greatest drawbacks was transport of bulky goods, such as cotton.

A military background may have fulfilled many of the requirements of an early political officer, but from the 1920s the Colonial Office sought to employ administrators trained more in administration and legal subjects, reflecting on the need for a greater practical understanding in this area. A special one year course was run at Oxford University to train such men. The Great Depression and fall in the price of cocoa put an end to greater dreams, but not before Guggisberg had founded Achimota College, Korle Bu Hospital and good port facilities. So high was his standing that some of the people of the Ashanti Region came to pay homage at his graveside in Kent, after his death in 1930, but found an unmarked grave. Shocked, they immediately erected a gravestone in his honour to remember him for his sincere developmental aims, and love of the Gold Coast people.

Another luminary was A. W. Cardinall who arrived in Ashanti in 1915, and was part of a team who merged German Mandated Togoland into the Gold Coast, and had the moniker "Bugin di Kugur'" or "fire has consumed the stone". He made a humorous remark in his diary of 1919, when asked by the Governor's wife to write an article "A day in our life on the Gold Coast". "I refrained, for if one was to write a faithful account of an average day here one would either appear as a modern day Solomon or an incompetent idiot. Not presuming yet to be the former, I had no inclination to proclaim myself the latter." That day he had three decisions to make: first about the ownership of a sacrificial place, second a debt needed to be repaid or a father would take back his daughter, and the third was the suspected sale of a wife. He was a budding ethnographer and wrote books about the North and Togoland. On retirement he also compiled the first census, giving useful details of the previous years. He went on to be Representative for Great Britain at the League of Nations, Judge and Commissioner in the Cayman Islands, then in 1941 Colonial Secretary and Governor of the Falklands. Charles missed meeting him by a few months.

It had taken almost 20 years to bring about law and order and then to establish peace in the Tong Hills (Tallensi area), and North Mamprusi generally. A book of excerpts from official diaries of early colonial officers, "The Gold Coast Diaries" by Thora Williamson, gave a glimpse of the life and times of the men in the Gold Coast between 1900 and 1919. The chapter on the North was especially interesting as Charles referred to those men as being called "Lions of the North". They had implemented direct rule and their decisions were in evidence everywhere in 1930; their personalities had become almost mythical. The change to indirect rule was about to be introduced, built on the decisions of that previous generation of men. Direct rule was barely established before the challenge of indirect rule was being contemplated by central government.

By 1930 basic law and order had been achieved in the Northern Territories, and progress was now needed. The few colonial officers working in the area were aiming to bring change quickly, and most had written a simple study of the people in their area. Fortes acknowledged the deep complexity of the culture composed of two distinct, but merging, groups. Their individual family compound differences were still of great importance to them, and the source of constant worry and quarrelling. The earlier raiding and enslavement of people was still fresh in their minds, exacerbated no doubt by the years of the Transatlantic slave trade. Many local markets had closed, and this discouraged surplus food production. A ready surplus would have encouraged more predatory behaviour from neighbouring tribes. The British were always rumoured to be on the point of departure, and the old order might return. There was a lack of any rudimentary scientific understanding, especially in the field of human and animal health. There was also a fatalistic approach-that life and all happenings was in the control of "wuni". In Northern Nigeria, by contrast, the shaduf, or manual water pump, was in use prior to the British arrival, along with Islamic state structures and writings.

Ancestor worship was still strongly adhered to and in 1930 governed all aspects of Tale life. To colonial officers it must sometimes have seemed like medieval thinking and witchcraft. Direct rule had raised the Mamprusi Na, and in the Zuarungu District the Kunaba, as Paramount Chiefs, whose ties were to the Dagomba Na, the Ya Na. The area chiefs, too, had nominated henchmen, the Kambonabas, who had since British rule become more powerful than the Tendanas who represented the original inhabitants. Colonial officers dealt almost entirely with the Chiefs, who greeted them on arrival in the village and provided rest houses, firewood and water. This one-sidedness created fear and insecurity in the minds of some of the people, that they might not be believed or treated equally. This made many of the people especially anxious to understand the wishes, and keep on the right side, of the DC or white man. The Tendanas had become somewhat overlooked. Indirect rule aimed to redress these unintended consequences.

The original inhabitants of the southern part of the Zuarungu District, the Tallensi, spoke a dialect of Mole-Dagbane. When asked to attribute their origins they said their founders emerged in various ways, such as from holes in the ground, the hollow of a tree or by descending from the sky, showing that only these descendants had rights to provide the earth priests or Tendanas. In 1930, the Tallensi went mainly naked or in a simple animal skin, carrying bows and arrows as their fathers had done. It is thought that some two centuries before the coming of the British, Mamprusi Namoos clans from Nalerigu began infiltrating the Tallensi, with the original settler Mosuor being buried in Tongo, centre of the Tale area. They had horses and went clothed, and viewed nakedness as beneath them. They were originally people of superior culture, of obscure origins, and may have been an offshoot of the Songhai Empire or another Sudanic Empire of the Niger bend. They gave mystical power and protection to the Tallensi, especially with essential rainmaking. They remained distinct; both groups celebrated their differences. For the Tallensi it was vigorous festivals such as the Gologo dances, while for the Mamprusi it was Dama and their naam, the sacred giving of a chiefship, and the relationship (in Tale, Mosour bii) to each other - who "follows" whom.

8

The coming of men with white skin, or "red" as it was perceived, the destruction of the Kingdom of Ashanti with superior military might, the downfall of freebooters like Samory and Babatu, and the return of slaves to their former homes must have been seen as amazing happenings in one lifetime. Stories must have circulated about the slave ships who took men away. Add to that the new phenomenon of locusts and rinderpest, which decimated their already frugal food supplies and destroyed their cattle wealth. What did these white people want? The incoming Mamprusi wanted chieftainship of the land; the slavers wanted people to sell; the traders traded beads, cloth etc. "Nasara Panga", white man's power and rule, brought a change that was incomprehensible in the local mind as to where it would lead. The white man wanted carriers to bring loads from the coast up to the North. Road makers and builders of government buildings and houses, all involving forced labour. Local people initially could see none of this being of relevance to them. Payment for food was given to the Chiefs, who seldom passed on the full amount. No other invader had done anything like this. The cowrie shell, a century's old currency used in trade across the Sahara, was replaced by metal coins. The old trade routes though diminished, were simple paths which led over the scarps down to the plains beyond, to Salaga or Ashanti. But the white man needed 14 ft- wide roads to run his strange objects of transport. His cars, motor bikes and bicycles needed to cross rivers, so drifts (shallow fords), like the one at Pwalagu, had to be reinstated after the rains each year. Local people had to do this laborious job under forced labour until 1936, but as late as 1925 the District Commissioner wrote to the Secretary of State for the Colonies that roads were seen, at that time, of no real economic advantage to, and by, the local people.

On seeing the first bicycle in the mid- 1920s the amazed Chief of Bari remarked that he had now seen "all the wonders of the world"; it was indeed termed an "iron horse". Gramophones, and later radios, were another marvel – how many people were inside the box? Only 30 years had brought remarkable dynamic change in all areas of life in North Mamprusi. This was the situation that Charles found when he arrived in 1930.

Two years after first arriving in the Gold Coast, Charles was posted in 1932 to Zuarungu in North Mamprusi, which was to be his base for almost all his 17 years in the Gold Coast. Zuarungu was 98 miles north of Tamale, and six miles from the important Tong Hills. An agricultural officer even visiting the far north was rare in 1931, and so his role was not defined. At just 23 years old Charles began his agricultural survey of the Zuarungu district, and introduced "mixed farming" to improve the nutritional standards of an area prone to famine, malnutrition and soil depletion. The main effort initially would be concentrated in the Zuarungu district, where six families were surveyed. The work was eventually to carry out a comprehensive agricultural survey of the current systems of farming, areas of land cultivated, varieties of crops grown, methods of storage and disposal, manpower available, marketing data, wild sources of food, livestock in the economy, water supplies, geology, soil and weather conditions, followed by field trials of possible improvements.

The Chief Commissioner strongly approved the programme. The survey was in line with the policy of indirect rule through the native administration, which was receiving much attention at this time. Factual information was badly needed. In 1932, nutritional improvement and economic development were essential in North Mamprusi, and mixed farming was hoped to be the solution. Local trade, such as it was, had been the basis of the economy, and agriculture was subsistence only. Some markets only sprang up in times of famine or hunger. Traditionally, the area was 90% dependent on grain production, but this was very susceptible to locust depredation and severe weather conditions. Any changes were viewed with suspicion, such as efforts to encourage more yam production. Traditionally, continuous agriculture was practiced on the compound farms, with shifting cultivation on the more distant or 'bush' farms. 98% of the population were farmers but only 20% were dedicated to farming. Around only 10% of the land was manured, causing low fertility and crop yields. The annual burning of grass and all ground vegetation added to the problem of general dryness, and encouraged soil erosion during the rains. It was believed that the ancestors had demanded the burning, and ceremonies included this activity. Charles thought slave raiding could be one reason as it was important to see the enemy approaching. The residue of the burnt grass would also provide some small short- term fertility. He managed effectively to reduce this burning, and to get people to use much of the grass for thatching, fodder and farmyard manure.

The Agricultural Department at that time had a low profile and was centred on Accra, focusing on cocoa and sapling trees of economic importance. With the Veterinary Department they were tackling the new scourge of rinderpest and locusts, a problem in Africa south of the Sahara since the turn of the 1880s. The First World War, and retrenchment during the financial crash of 1929, led to a slowing of development, loss of essential personnel and emphasis on export crops. A Department of Agriculture report mentions frequent food shortages every two to three years in the Zuarungu District. A visiting dietician from Nigeria advised the Chief Commissioner to request an agricultural officer to examine the reasons for famine and malnutrition especially amongst the Tallensi (also nicknamed Fra-fra), in an area of 780 sq miles. Bad seasons mostly brought hunger, not starvation. In the last drought people were said to have been reduced to eating grass, and in the time before British rule some had sold their children into slavery in times of famine. For every 100 babies born alive in the Northern Territories in 1930, 47 died before puberty. Famine, soil depletion and disease had left this overpopulated area in distress. The population was approaching 200 per sq mile, with an adult cultivating a little over two and a half acres. With the introduction of a simple plough or donkey cultivator this could be increased to eight acres involving only one man and a boy. The agricultural bulletin written by Charles in 1937, "Agriculture in North Mamprusi", is full of details, but how it came to be written and the way of life at that time is not recorded. The challenge for Charles was enormous.

There was a negative attitude in the Agricultural Department in Accra where staff felt, despite their best efforts, that nothing commercially successful had come from Tamale Research Station (South Mamprusi) or could come from the North. Its poor roads and lack of facilities gave it a low priority. Roving agricultural officers only reported the situation. One of the reasons for this underdevelopment lay primarily in the lack of finance, hence the eventual introduction firstly of the Messenger Tax. This would lead to the Compound Tax and various taxes around market activities, and so forth. Even in 1935 the then Under- Secretary of State for the Colonies, paying the area a visit, was amazed to learn that there were only two agricultural officers north of Kumasi.

In 1934, a year of momentous change, indirect rule was being put into practice and Native Authorities created. Charles was subordinate to the political control of the Gambaga District Officer, who at this time was George "Gumbo" Gibbs. Gibbs arrived in 1933 from duties in the south, and was also responsible for the Zuarungu District. This was because retrenchment had left the post unfilled. He quickly saw the value of an agricultural emphasis to mitigate the recurring famines, malnutrition, mineral deficiency, soil erosion, and so forth.

In January 1934, the anthropologists Dr and Mrs Meyer Fortes arrived in the Zuarungu District. In August, the political administration had a new acting DC in Zuarungu, Allan Kerr, and a new reforming Chief Commissioner, William "Kibi" Jones (later Sir) in Tamale. Fortes's aim was to study the Tallensi within the framework of "Structural Functionalism" and to make anthropology a true social science, or social algebra as Charles would have remarked! Very quickly he told the administrative officers the job they were undertaking was much more complicated and difficult than they could imagine.

"Our reconnaissance of the cultural landscape of the Tallensi has left us with a tangle of facts which do not fall into an obvious pattern. The reader may rest assured that his perplexity bears no comparison to that of an ethnographer seeking, for the first time, to understand Tale culture and social organisation. The perspective changes as the investigator shifts his vantage point. The greatest puzzle of all is why there is so little apparent unity and cohesion in an area of such dense and close habitation, carrying a population that has been sedentary for a long time, has a more or less homogenous culture, and a strong sense of the continuity and stability of its social order." (The Dynamics ofClanship amongst the Tallensi, Meyer Fortes, 1945, p.27).

No wonder the administration found it difficult to deal with. In order to do his fieldwork, Fortes asked to be included in all details of local life. The Chief of Tongo, immediately suspicious, asked if Fortes wanted to report back to DC Gibbs or pass judgement in disputes. Later they sang songs at the Gologo dances, saying Fortes had come to live amongst them, and now had stolen their secrets. The Compound Tax and later Hut Tax led one man to ask Dr Fortes what it was for and why they had to pay it when the British had their gold mines and banks. It was, of course, explained to them many times over. The idea of any form of local government, such as indirect rule with the formation of a native administration, was a foreign- imposed idea. DC Allan Kerr's diaries show he continuously explained court tribunal procedures, running of the Native Treasuries and the formation of the earliest form of local government by the Native Authorities.

In Fortes the Tallensi found a voice to recount their concerns and grievances, sometimes asking him to mediate with the authorities; however, he would relay the facts and try to remain impartial. But he too worked primarily with the Chief's section of the population. The people could talk to Fortes in Tale, as they never could the DC. The court interpreters, especially Bassana in the Zuarungu area, gained great control and tribute via this role. Fortes recalls in his diary that he was asked be a personal interpreter in very secret sensitive matters with the Chief of Tongo, the Tongrana. So too, of greater growing consequence, were Christianity and Islam, threatening their cult of ancestor worship and the very basis of their ideals and societal beliefs.

Indirect rule sought a change to create a chain of command, for all the community, not originally found amongst the egalitarian Tallensi people. It sought a return to certain supposed former traditional ways, and Fortes looked critically at all colonial officers in his area, as well as the locals he studied. Criticism of the role of the District Commissioners by the incoming "specialists", and anthropologists like Rattray and Fortes, picked up on the supposed lack of development before 1933. The notion "that up to 1933 the policy there was one of inactivity and of letting well alone" was aired in the West African Review of July 1939. It caused the old lion Walker Leigh, now retired to Devon, to reply "I take exception to that statement". He pointed out that in the early days the people were not easy to govern, that due to the discipline of the early pioneers the present changes could be made. "... their predecessors, who changed a hostile population into one of the most amenable and orderly people in Africa and this was not done in the Office." Talking of the hardships and means of travel he remarked, "It can be said that there were no joy riding visitors, when it meant marching from Cape Coast to Gambaga, quite a pleasant journey in the rains, with a lot of swimming for those who were good at that sport". He ended with another dig, remarking that for them their "... time had of necessity largely to be spent in outdoor work rather than doing clerk's work in an office, or delving into Native Customs of the past". There were almost no literacy amongst the northern tribes, and clerks and other essential personnel were brought up from the South.

Fortes firstly created unease for the colonial officers, but suspicion changed quickly to respect. With acceptance and confidence in his abilities and good intentions, "Kibi" Jones and Gibbs were able to formulate policy with a better understanding of the local situation. Time was short, and progress crucial. Fortes wanted his work to be honestly relevant, but had no direct say in governance; Jones and Gibbs had the responsibility and wanted to understand the situation to form the best outcome. Conflict of interest was averted by co - operation and respect. Jones asked Charles once if Fortes could be "trusted", and was again reassured. DC Kerr once summed up the move to indirect rule as to "give instruction NOT instructions". There was a certain amount of resistance on all fronts to what seemed continual sweeping change, particularly from the older generation, and some people boasted of never having seen a white man – and did not want to either. Ideas were often treated with suspicion, and it was the personal interaction with the colonial officers that led to personal confidence and acceptance. DC Kerr in September 1936 visited the old dying chief of Pelungu, who at that time was one of three surviving chiefs prior to British rule; Pelungnaba asked him, on his behalf, to be remembered to the King. He remarked that but for the British he would long ago have been dead from a poisoned arrow. Peace, law and order were important.

Even in 1932, Charles had to trek with carriers along narrow paths about 18" wide in all weathers, usually up to 20 days a month. All necessities were carried on the head of carriers with loads of up to 60 lbs. Europeans were few and far between, but the bonds between them were strong and messengers linked them by notes sent in a cleft stick, and sent post to post. From 1938 the Agricultural Department in the North grew steadily, and the three newly created research stations had freshly appointed agricultural officers supervised by Charles. At the King's Birthday ceremony in 1943 he received the MBE for services to agriculture, the same year as "Kibi" Jones, the Chief Commissioner, got a knighthood. By 1947 Charles was ready to move on after founding a new agricultural research station, now a Natural Resources College at Nyankpala near Tamale.

Fortes placed all his personal diaries and papers from his fieldwork in the Cambridge University Library. In one letter he says he wishes his study area to be "a part which is not a complete backwash, but is of real interest and coming importance to the Administration so that our work will not be left to languish in the British Museum, but will be of use to those concerned with the development of Africa". In 1964, Professor Fortes, on another field trip to Tongo, wrote to Charles and remarked that amongst the Tallensi they were legends in their lifetime. Fortes was asked about many colonial officers, and the word "devoted" was often mentioned.

I have used Charles's place name spellings, but three maps from the 1930s show differences amongst them. Today's spellings can be quite different again. Further difficulties arise as the Tallensi often drop a final vowel if not semantically significant. Fortes does this in his diaries, hence the different spelling.

In writing and researching for this book, I really understood why Charles found his time as Director of Agriculture, Northern Rhodesia, aged 44, such a total contrast, and at times so difficult. From his second tour onwards in late 1931, at the age of 23, his life was spent in a totally African environment. Sometimes he would go many days, or even a week or more, without seeing another European. His work focused only on indigenous people, their wellbeing and education. In the 1950s, Rhodesian federal development focused predominantly on white settlers, and their style of life.

Northern Rhodesia was a Protectorate like the Northern Territories of the Gold Coast, but there the local similarities stopped. It was politically a very different country, even from its neighbour, self-governing Southern Rhodesia. The 10 years of federation from 1953 led to many government departments being split into European and Territorial.

Charles kept daily Diaries, but Marjorie burnt them to forget the memories of those stressful times. 1950s Lusaka was very much a developing European town in style and layout. European families of all generations lived there and owned land and houses. This was in total contrast to the Gold Coast. Charles had a career governed by the political changes of both indirect rule and federation. But rising nationalism was to make both fail in the long run; political times in Africa were moving too fast. The contrast in the field of agriculture between them was also enormous, too. The famines and poor soils of Northern Ghana were in significant contrast to the relative fertility of Zambia. Charles always maintained that given the right situations,

Zambia could be the food basket for Africa, hence his interest in the development of the Kafue Pilot Polder.

I was well on the way to writing "The Long Garden Master" when in late 2008 Dr Moses Anafu of Ghana returned to England and came to Cambridge. It was timely as his help and research has greatly improved this book. In 1934, by chance, his father was among the men of Bari who re-roofed Charles's house. He showed great personal interest and referred to colonial officials I knew well by their initials. He pointed out to me that my father's work, and that of his colleagues, was part of his tribal history in the same way it was part of mine. I must thank too, the Cambridge University Library, which is a treasure trove of information, and houses the Fortes papers.

My husband Victor Meaden has been a helpful and patient supporter. He has done wonderful work on my portfolio of fading photographs from the Gold Coast. He has patiently seen our house taken over by books and papers for too long, as I have slowly progressed.

Sylvia Lynn,
St John's Road
Cambridge 2012

Chapter 1 - 1908–31

Early Days

I count myself as fortunate in always knowing that I wanted to be associated with agriculture. Early impressions gained on my grandfather's farm at Haxey, Lincolnshire, had left a deep mark. I owe much to my mother as my father was killed in the First World War in December 1916, when I was eight and my brother was six. I was lucky in the boarding school to which I was sent at Ingatestone, a typical Essex village, where most of the day boys had farming backgrounds. After leaving school in May 1924, I learned the joy of working in the field, and the exciting quiet of the early morning. All the fieldwork in those days was done with horses; tractors were only just coming in.

In October 1924, I went to Wye Agricultural College (1) with a scholarship from Kent County Council to take a diploma course; there I spent three happy years, which set the course of my life. Wye is beautifully situated under the North Downs, on the edge of the Weald of Kent, surrounded by intensive farming including hops and fruit as well as livestock and arable cultivation.

A special excitement at Wye in the summer of 1926 was to visit Canada on a harvesting trip with 40 other students under the auspices of the Canadian Pacific Railway. We were located on farms in various parts of Saskatchewan, through the good offices of the Women's British Immigration League. It was good experience with the joy of working physically and feeling more tired at midday than one thought possible, with half a day's work still to come.

Towards the end of the summer term in 1927, the big question arose as to what one would do when one had achieved a degree or diploma. The outlook for farming in Britain was bleak at that time with few openings for Assistant County Agricultural Organisers, Estate Agents or Farm Managers. The alternative was to seek a post abroad.

The Colonial Office was enlarging the Colonial Agricultural Service, (2) and there were opportunities for securing a student grant from them or the Empire Cotton Growing Corporation. I applied for interviews with both and was offered and accepted a studentship from the latter. The Cotton Corporation derived its funds from a cess or tax on every bale of cotton imported into the UK. Its purpose was to train staff for research in order to increase cotton production in the Empire to replace cotton previously imported mainly from the USA.

The studentship took me up to Cambridge for a Postgraduate year at St. John's College. We spent a happy year with Mr Engledow (later Professor Sir Frank Engledow, Drapers Professor of Agriculture) as our supervisor and Guillebaud as tutor. The number of lectures was not as heavy as it had been at Wye; we had more freedom with time to use the library. We worked on Mr Engledow's experiments on Cage Field on the University farm and spent some time during the winter counting tillers for the census of an acre of wheat. At Cambridge a scientific approach was added to the basic knowledge we had acquired at Wye; we studied Statistical Method, National aspects of Agriculture, Agricultural Economics, Systematic Botany, Tropical Agriculture and Genetics. The Survey Unit in the Cambridge O.T.C. taught us to use a theodolite.

After my year at Cambridge I went out to Trinidad, for the second year of my studentship. I travelled on the Elder Fyffe banana boat "Carare" to the West Indies to study at the Imperial College of Tropical Agriculture in St Augustine, Trinidad. (3) The island was chosen as the site of the College because of the variety of tropical crops which could be grown there. It was already the headquarters of the Imperial Department of Agriculture of the West Indies with a great reputation.

In Trinidad, we acquired a tropical finish to our basic knowledge of agricultural science, and learnt something of the content of life in the tropics. Life at the College represented for most of us a completely new way of living. At nightfall the darkness was filled with the noise of cicadas and bullfrogs, and lighted by the emissions of fireflies. During the day, lizards of all sizes and colours chased insects in the sun and we made the acquaintance of chameleons and humming birds, little bigger than insects, hovering and darting in and out of hibiscus flowers.

For an award of the Associateship of the College we were required to carry out an individual research project and submit a thesis as well as pass examinations. I explored the floral mechanism of the groundnut. My research was tampered with in a student prank, and I was asked to submit a paper to qualify on my first appointment. In 1937, I fulfilled this with the publication of "Agriculture in North Mamprusi".

We made excursions once a week to estates and other centres of agricultural interest to see sugar, cacao, citrus, pineapple and copra. We studied land settlement schemes, tenantries and Government Research stations to fit in with the subjects of our lectures. Everywhere the reputation of the College was high, and we were well received. Our academic year in Trinidad was about long enough to achieve its purpose, and by this time we were anxious to know to which territory we would be allocated. By the terms of my studentship it was incumbent upon me to seek a post in a cotton growing country.

Since there were no suitable appointments with the Empire Cotton Growing Corporation I was referred to the Colonial Office and saw Sir Frank Stockdale, the Agricultural Adviser to the Secretary of State. (4) Subject to passing a medical examination I was offered an appointment on probation, by the Secretary of State for the Colonies, as an Assistant Superintendent of Agriculture in the Gold Coast, on a salary of £480 per annum. If confirmed in my appointment after three years my salary would rise by annual increments of £40 a year to a maximum of £920. I was delighted and accepted the offer; a job in 1929 was worth having especially one that was permanent and pensionable.

A great advantage of service in West Africa over East Africa was a slightly higher salary and more frequent leave. We were given one week's leave for each month served, a normal tour being 18 months, generous but necessary from a health point of view. Because of my age I had to wait in England until I was twenty-one and a half in December 1929 before taking up my appointment.

I received a kit allowance of £60 from the Crown Agents, and an advance of salary to purchase a motor car. I secured a Bullnosed Morris Cowley, a 1926 model for £45. The cost of shipping to the

Gold Coast by Elder Dempsters for a passenger's car was £7.10s. Acquiring basic equipment to set up an establishment was exciting; I learnt that everything had to be taken out- very little in the way of household or camping equipment could be obtained there but Government did supply housing and heavy furniture. The superb service of the tropical outfitters Griffiths & McAlester must not go without mention.

They made life possible in West Africa for expatriates in those days. I handed over my car to them in Liverpool for shipment, and they placed on board my various purchases, which included my camp equipment, mosquito boots (5) and uniform cases, crockery, cutlery and cook-box. I added some board games, recorder, ukulele, gramophone and 75 rpm records, saddlery and tennis racket. My deckchair was to be available while on board. I also acquired two of their splendid chop boxes (6) which contained a small supply of basics sufficient to last for two or three weeks, and a few luxuries which all proved of great value in remote rest houses in the months ahead. They also sent out papers regularly. They could be relied upon to remember special days for those at home by sending suitably selected chocolates or flowers to be debited to one's account. Payment was painless by monthly Bankers order during the course of the ensuing tour.

On 4th December 1929, I took the good ship M.V. Aba (7) of the Elder Dempster Steamship Company, and sailed from Liverpool on a murky afternoon bound for West Africa. There was the usual bustle before the ship sailed, old friends greeting one another, telegrams being received. There were various rumours circulating around the ship before we left; she was one of the early turbine ships and had recently been converted from cargo status to taking passengers as well. During alterations, concrete was put down along her keel to give stability; little did we know that this was going to be tested considerably during the course of the next few days. At dinner that night, the fiddle boards were up on the dining tables, the boat was moving a bit as we got out into the estuary and the wind was rising. This was to increase so much that by midnight it was necessary to close the bulkheads. In the morning we found we had hit a 100 mph gale during the night off Holyhead causing damage to our steering and the port engine.

The Berengaria was one of the first ships to pick up our SOS call and my friend Norman Crockhart, sailing on her, telegraphed his mother to use discretion in informing mine. This buffeting continued for about five days before we had any idea of what was really going to happen. The sound of the wind through the superstructure was a continuous whine, and the banging of the waves on the side of the ship, completely submerging our porthole, was at times like thunder. From the Saturday after sailing, we were able to use the lounge with the chairs roped together, but we were not allowed on deck and the bulkheads remained closed. Apparently, on three occasions we went beyond the theoretical angle of recovery. Finally, the Dutch tug "Swartzig" came out from Swansea to take us in tow, six days after we had left Liverpool. She towed us for the next two days successfully into Queenstown Harbour, by which time the seas had moderated.

The engineers eventually reported that our engines could be repaired, and we continued our journey on 16th December from Queenstown to Takoradi in extremely pleasant conditions. At the Canaries the ship's company donned their white uniforms, and male passengers blossomed out at dinner in white mess jackets, which were so much more comfortable in the tropics than dinner jackets. These also provided an opportunity to show off cummerbunds in territorial colours, old gold for Gold Coast, dark green for Nigeria and red for the Royal West African Frontier Force. (8) Passengers came together like long lost friends.

On the 29th December, 25 days after sailing from Liverpool, we came, in the early morning, into Takoradi Harbour bringing the Christmas mail. Like all the west coast ports, Takoradi had the characteristic acid/sweet smell of fermented cacao beans and palm kernels but was a clean- looking modern port which had been opened by the Secretary of State for the Colonies just a year before. I was informed that I was not to disembark at Takoradi but to go on to Accra, the capital, another six hours sailing up the coast. It was interesting to see the low coast, with the heavy lines of white surf all the way, and low hills covered in dense vegetation in the background. At regular intervals, showing up white in the sun, there were stone castles of European origin, built variously by the British, Germans, Danes, Norwegians and Portuguese, grim reminders of the slave trade. During the evening we saw three waterspouts that came quite close to the boat as they twirled together for a time like dancers; one could hear the water beating on the sea.

Landing at Accra the following morning was a thrilling experience. The ship anchored in the roads half a mile off shore and a wooden "mammy chair", holding four passengers, was lifted from the deck by the ship's derricks and deposited in a surfboat alongside. Passengers seated themselves in the surfboat and the mammy chair was hauled back aboard. Muscular paddlers, six on each side, with rhythmic song propelled the boat with great skill towards the shore finally choosing the right wave to carry it well up the sandy beach. Cars were carried on two surf-boats lashed together; I was glad not to see mine landed!

First impressions were favourable as the season was the harmattan when a drying wind blows across from the Sahara Desert in the north, causing low humidity and spectacular sunsets from dust particles in the atmosphere. It was exciting to unpack various loads, which had been disembarked that morning, including my new camp bed, which was promptly erected complete with mosquito rods and net on the veranda. I learned that I was to be posted temporarily to the Division of Meteorology & Publications, with a desk in the library in Headquarters on Rowe Road, a large wooden building opposite the Police Headquarters.

My first weekend was spent in Aburi Botanic Gardens, (9) a hill station a few hundred feet up in the rain forest 30 miles behind Accra; it was beautifully cool at night making a blanket welcome with a Scotch mist falling in the early hours. The trees were large and impressive, especially the Royal Palm Avenue leading to the Governor's Rest House. Aburi was also the headquarters of some of the specialist services of the Department of Agriculture, the Mycologist and Entomologist being stationed there. The vegetation was tropical rain forest, typical of the cacao country, which stretched inland for over 100 miles to Kumasi and beyond. Before the First World War the Chief Curator, until renamed Director of Agriculture, lived here. But agricultural officers were still known as Garden Masters.

One of my first jobs was to invigilate an entrance examination for Cadbury Hall, the departmental training centre for African staff in Kumasi financed by the famous chocolate firm. They ran 3- year agricultural and forestry courses. The Director, G.G. Auchinleck, suggested that I should take every opportunity to go on tour with any officers visiting outstations. One such trip was to the sisal plantation outside Accra, beyond Korle Bu, (10) the big hospital.

The emphasis was always upon cash export crops of which cocoa, of the order of 220,000 tons per annum, was far and away the most important. The bulk of the staff was concerned with the preparation, handling and marketing of cocoa rather than with its production. It was 10 years later before the Cacao Research Station was opened at Tafo. Additional cash crops were palm oil and palm kernels, coffee, cotton, copra, citrus, bananas and rice. The local food crops were maize, cassava, citrus, sweet potatoes, groundnuts, guinea corn, pigeon peas and cowpeas they received some attention, although there was insufficient staff to do much with these in 1930.

Livestock played a relatively small part in agriculture, and animal husbandry linked to animal health was mainly the responsibility of the Veterinary Department, which was active in the Northern Territories and on the Accra Plains. The place of livestock in the economy was to change considerably 10 years later when the Gold Coast was drawn into the War. Then food crops and livestock became all important, and the value of the work of the Agricultural Department was immediately apparent.

Life in the Gold Coast seemed much cut off from the rest of the world. The long sea journey from Britain; the dramatic landing in surf-boats; the use of mosquito nets and mosquito boots; the boiling and filtering of drinking water; the daily prophylactic dose of five grains of quinine against malaria; the absence of European children and adults over the age of 55 years; the wearing of Wolseley helmets in the sun between the hours of eight in the morning and four in the afternoon, all served to remind one that it was a special place where one should not take liberties. (11) There was no radio in those days, and no refrigerators, though blocks of ice were distributed daily in Accra and Kumasi. The fortnightly mail from home and one's books and papers were all important, as well as social clubs in larger centres which included a library and sports facilities. There was a good deal of home entertaining, particularly by more senior officers.

The Gold Coast house servant was very efficient and proud of his position, and had a wonderful sense of humour. Pidgin English was spoken to houseboys who would "pass chop" or "pass a bath". Carefully cut oranges, very juicy and refreshing, were served with tea and appropriate clothes were put out for the day or evening. Games were played with enthusiasm, tennis and polo being popular. Inevitably a lot of "shop" was talked because individuals were keen and dedicated to their work. There was also a dearth of white feminine company to divert conversation, so talk consisted very much of the highlights of the last leave, and the prospects for the next leave.

After a month I was posted to Asuansi, (12) an Investigational Station in the Central Province about 16 miles behind Cape Coast. A lorry from the Government Transport Department picked up my loads, and with my cook and steward boy I made my way by car through Nsawam, Swedru and Saltpond almost to Cape Coast. Then inland to Asuansi, a distance of 60 miles from Accra. I took over the station from an old Wye man, Marden Cook, who was being transferred to the Government coconut experiment station at Atuabo, in the Western Province. Marden was a colourful figure and a fine raconteur; his conversation being worthy of Blackwood's Magazine. It was wonderful to find myself in charge even temporarily of a station of about 200 acres, well stocked with all kinds of forest crops, cacao, coffee, oil palms, coconuts, and a small number of rubber trees. The fruit trees were citrus, avocados, Cavendish bananas, paw-paws, and a number of plots of food crops such as maize, sweet potatoes, groundnuts and cassava made it a veritable tropical paradise.

It said much for my African staff that they bore so well with my immaturity and ignorance of local affairs. The purpose of the station was to serve as an agricultural base, with a few experiments of an observational nature. There was a nursery to raise planting material for distribution, coconuts being germinated here and sent all over the province. The garden around the bungalow at Asuansi was well planted, by men whose basic training had been obtained at Kew Gardens. All the ornamentals were labelled with their botanical and common names. There was a magnificent Royal Palm Avenue leading from the house to the office, a quarter of a mile away, altogether a superb introduction to a tropical flora.

My steward boy Jacob (13) was a Kru boy from Cape Palmas up the coast. He had come to the Gold Coast, like many others, to seek employment, and was a good servant and a pleasant man. He was particularly fascinated by my gramophone, a hand- operated H.M.V. portable, which he referred to as "them plate where plenty people sing for your country", and the "Anvil Chorus" was one of his favourites; also my calypso records from Trinidad appealed to him. It was particularly quiet in the forest at Asuansi, but occasionally the nights were disturbed by the screeching of the tree bear or Hyrax-

Dendrohyrax Dorsalis, (14) which nobody had warned me about, quite terrifying when first heard.

Towards the end of March 1930, I was told I was to be transferred to the Northern Territories to be stationed at the provincial town of Tamale.The railway journey from Sekondi to Kumasi (15) proved an exciting experience. There was a milling throng of people at Sekondi Station as I put my car on a goods truck to come up next day. Then I caught the scheduled 9.04 am passenger train, arriving at Kumasi 120 miles away at 5.3O that evening, having stopped at every station on the way. Everywhere there were happy noisy people looking well fed and well clothed and accepting the railway as part of their way of life. It was a comfortable slow journey. I had a first class carriage to myself, shared with my newly acquired terrier Bill, the windows were tinted and a large fan kept the air moving. We travelled through bush country, with a few cacao farms and primary or secondary forest on either side. The line linked the big mining towns Tarkwa, (manganese) and Dunkwa and Obuasi (gold mining towns). My steward boy Jacob came through the train from time to time to serve drinks and an excellent lunch consisting of a pork pie from the Ice Company in Sekondi.

In Kumasi, I prepared for the journey north. This involved shopping in various stores, Pickering & Bertroud, the Swiss African Trading Company and the French Company. It was necessary to take up a good supply of stores. People stationed in the Northern Territories sent their orders to Kumasi, and these were despatched through the Government Transport Department on the weekly mail lorry. There were also no banks north of Kumasi.

In due course, my loads were packed into a Government lorry for transport north to Tamale, (16) and I set off with my steward boy in my car. I was proposing to break my journey at the Volta River 140 miles north and stay in the Yeji Rest House. The tarred road ended a few miles out of Kumasi, and the road then changed to laterite gravel which became heavily corrugated. One had to drive at over 40 mph in order to overcome the corrugations, with the danger of shaking a light car to pieces.

There were towns at Mampong, where we left the forest behind, Ejura, Atebubu and Prang along the way but otherwise little signs of life, water supplies being restricted through this sandstone country. The road was dusty, and lorries generally stopped for private cars.

The Morris stood up well and in due course we reached Yeji tired and dusty; darkness fell but there was no sign of my lorry- subsequently I learnt they had had trouble on the road. Typical of West African hospitality, the tsetse fly research doctor, Dr Jackson, had a mosquito- proofed room in his bush house at Yeji. He split his camp bed, and I slept on the canvas frame whilst he slept on the mattress on the floor.

Going North for the first time 1930
Bullnosed Morris and Government. Lorry

" The Beetle", T. Lloyd Williams,
and Charles 1937

Captain Jock Stewart, Director
Veterinary Services, and Charles
1932

Alhadji Almajiri Mahamadu, Hausa teacher, Tamale 1931

The Cricket Team 1930.(left to right) Fraser (Police) Symonds, Oakley,
Cowan, Lynn, Miller, Wentworth, Candler. 2nd row Stewart, Lokko,
Hamilton, Cockey with African Staff players

Chapter 2 - 1st Tour 1930–31

Life in Tamale and Locust Patrol. aged 22

The Volta River (17) was wide and swift flowing at Yeji. The excellent motorised ferry took two vehicles and many passengers across at a time. The approaches to the river traversed typical riverine vegetation, and the water was discoloured by silt topsoil from up country in suspension. The river formed the boundary between Ashanti and the NorthernTerritories. (18)

The road north stretched for 100 miles to Tamale, with only Salaga a town of any size between. I was given a house on the research station, where I was to work, from where I wrote the first of many letters home to my mother and brother.

9th April 1930 "My dear, dear folk,

I am sitting on the plinth in front of the bungalow labelled A.P.S. of A & F in which title I rejoice during my residence here. That is Assistant Provincial Superintendent of Agriculture & Forestry. It is a gorgeous night; I wish I only waited for you both to come along to dinner. If I describe myself as I am now you will have a typical evening.

I have been parking cards (19) again this evening and have just finished calling. It is an awful farce waiting till everybody is at polo, and then shooting in with a couple of cards. I returned to find two people had returned my calls – it seems so funny having people calling on me.

Chairs were out on the plinth – a nice raised square in front of the bungalow, small coffee table out with well polished brass tray, glasses and the usual drinks – whisky, gin, ginger beer, lime juice, vermouth, bitters and soda. It is amazing how quickly one settles down to the life though.

It is cooler tonight than usual, a brightish moon, high stars and a faint breeze, all very peaceful, except for the chirping of frogs and crickets. We had a hyena round the other night – Bill will have to take care. I believe they will come into a bungalow after a dog. Bill sleeps under my bed in the mosquito- proof room, which is such a joy as I need no net.

We start at about 6 am, with a break from 8–9 for breakfast and then carry on until 1 or 2 pm and then finish….".

The agricultural station of 300 acres was located two miles up the north road out of Tamale. This was the agricultural base serving the Northern Territories, the bulk of which is located on Voltaian sandstone stretching through Dagomba and out to Bole to the west, Salaga to the south, Yendi to the east and Mamprusi to the north. On sandstone the people lived in towns and villages scattered in relation to water supplies, and walked out to their farms in the bush outside.

Tamale (20) was the headquarters of the Northern Territories Protectorate, the Chief Commissioner being based here and the provincial heads of departments– agriculture, education, health, police and public works; the Director of Veterinary Services was also stationed here but was soon to move to new headquarters at Pong Tamale 18 miles north. Major Jackson was Chief Commissioner, Northern Territories, when I arrived; I just missed Major Walker-Leigh, said to be the last of the Lions of the North, (21) who had left a wonderful reputation behind him. John Symond was the Provincial Agricultural Officer; he was the only agriculturist in the North at the time. Gordon Cowan, the other agricultural officer, was on leave in Scotland.

There was a flourishing polo club supported particularly by the Regiment. The parade ground in the Cantonments made an excellent polo ground. The army supplied fatigue parties to look after the goals and sidelines. Every European in Tamale was required to play, and many that had never even thought of riding found themselves enjoying it. Mounted paper chases were a feature of Sunday mornings, generally at least one every month. Two mounted hares would lay a trail and a field of around 20 would gallop for five or ten miles before a very welcome shandy and breakfast put up in either a private house, the Regimental Mess or the Residency.

There was also a well- supported tennis section attached to the general club, one court being cement, the other beaten anthill earth. The Police Band, under an excellent Band Sergeant, played in the evenings on Sundays and Wednesdays – mostly selections from Gilbert and Sullivan and dance tunes. Many a pair of mosquito boots were worn out on the concrete tennis court dancing with the nursing sister, lady education officer (22) or station wives. There was a useful library at the Club. One was expected to attend on band nights; soon after I arrived I failed to turn up one night, and the next morning Dr Oakley, (23) the Assistant Director of Medical Services, was around to see if I was unwell. A cricket club was active, and provided an opportunity for Europeans to get together with the African community. The members of the African Club were mainly the clerical and junior staff from the various offices, mostly from the south.

Shooting was a popular pastime at weekends; one could go out locally and find duiker, oribi, guinea fowl or bush fowl, whilst further afield hartebeest, roan antelope, bush cow, water buck, reed buck, bushbuck, warthog and in some places elephant. Lion and leopard were not common, but they could be found by those keen to find them. Hyenas were abundant and made a hideous noise in the rains, when they could sneak up near to houses.

We were required to pass the lower standard in an approved language after three years, before confirmation of appointment. It was always a problem to know which language to learn. I took Hausa, which although not a language of the Gold Coast, was a lingua franca in all parts of the country. It was a pleasing language to the ear with a good literature. A Mallam, Al Hajji Almajiri Mahamadu, (24) came to my house; he had made the pilgrimage to Mecca on foot across the Sahara, taking seven years and gathering much wisdom and many skills on the way.

I enjoyed life in Tamale running the experiments on the agricultural station, until swarms of adult locusts (25) were reported to be flying over our northern frontier from the French Haute Volta region. These adults were laying eggs on their way south. The resultant hoppers, as the larval locusts were called, endangered the maturing guinea corn and millet crops with the prospect of food shortages later. The swarms were dense and reported to be taking four days to cross the frontier. This required emergency measures.

4th May 1930 Tamale "My dear Mother,
Cook the entomologist arrived yesterday and had lunch with me today; he is on the locust palaver. A labourer caught a rabbit this morning and gave

it to my cook saying it be good chop for white man. My cook jugged it and it was very good indeed. This week I have been having many battles with locusts. We are now going to try poison bait, besides the snag of poisoning stock, poisoning of natives will have to be considered.

Apparently, after a locust drive on the station the labourers rake over the ashes and eat the locusts like shrimps. Ugh! It makes me shudder. Yesterday, the sky was covered with a dense cloud of locusts which blackened everything under it. The position is likely to be serious with young maize and guinea corn coming on...."

All this happened at the height of the rains. Because of flooding the north road was open for only 20 miles beyond Tamale, at the Nabogo ferry. I was taken there by car, crossed the river in a dugout canoe and was met on the other side by two police orderlies. The overseer Atchulo, from the agricultural station, had assembled 21 carriers and a bicycle. We had only four miles to go to Diari where I spent my first night in a rest house, a mud-walled thatch- roofed building which smelt of goats, with open doorways and windows but made tolerably comfortable with my camp furniture.

The following day, I walked or rode a bicycle up the north road to the next rest house at Pigu, eight miles away, where I was put up for the night. I went through the usual drill of meeting the Chief and his headmen, and asked for news of locusts in the area; as he had nothing to report I went on next day to Nasia. This involved crossing the Volta River which was a very much bigger crossing than at Nabogo; it took quite a time to ferry the loads plus the bicycle in a dugout canoe across the fast moving river. It necessitated paddling a long way upstream on one side in order to hit the bank at the right place on the other side. The ferrymen were very skilful.

We trekked up the north road, about 12 miles to the town of Walwale, which was quite an important junction. Here the road went off east to Gambaga, the district Headquarters 30 miles away, and north to the Volta River crossing at Pwalagu. At Walwale there was one swarm of locusts which required attention. Whilst here I searched in vain for a missing hundredweight drum of sodium arsenate. This had been sent up on a lorry from Tamale before the road closed. It was to have been stored in Gambaga but it was reported not to have arrived.

Easter Sunday 19th May 1930 On Locust Patrol "Mother dear,
We returned this morning as per schedule from Walwale, and it is a good

spot indeed, I simply loved it. The rest house is built of mud, very large – a room at each end, chop room in the middle and veranda all round,the roof being thatch. The boys had put out the beds, and were ready to serve dinner when we arrived on Friday night.

Early on Saturday morning the Chief of Walwale, (26) a fine old chap, came up with two headmen and his sons – five naked little boys, bearing fresh milk and eggs. We gave him a dash of a bob or so, which he handed to his treasurer, and then tottered off. We took the rifles and guns at about 7 am two miles along the Gambaga road, and beat into the bush. I had a gun and stalked a two covey of bush fowls, but got nothing with two rounds. We returned for breakfast about 9 am.

A nice slack morning followed by a sleep in the afternoon. In the morning Williams and I were in the village looking round, and the Chief wanted permission for the band to play since a man had died. We gave our assent, and told him he could come and play 'small' for us at the rest house at 5 o'clock. We sat on the plinth and saw an army approaching from the village, well over 100 people, the old chief in front with his cabinet followed by the band, and wives of the dead man, shieeling like blazers. They came and danced for us, the chief drummer looked an absolute devil, pouring with perspiration half kneeling, shooting out an arm and shouting Na-a-a (are you well?). We know where the Black Bottom and Charleston came from. They loved it when I "Charlestoned" for them, and the band blazed up afresh; two fellows with bows, arrows and an axe shot about. We gave them a dash when we were tired of it, and sent them away with much Na-a-a and Toe-oe-oe, one of the Hausa greetings. It really seemed like Africa…."

My next tour was to Nalerigu to the east of Gambaga, an important town, and the seat of the Nayiri, the Paramount Chief of the Mamprusi. (27) I met the Nayiri on the road, at the boundary outside his village. He had his retinue of drummers and spittoon bearers with him. He sent out his Chiefs and headmen for news of the incidence of the locusts.

16th May 1930 "Dear Mother,

On Thursday, I had to go out to Tali, a village 18 miles away, where a swarm of small black hoppers were reported. I took a couple of overseers and stacks of poison bait, but found that instead of being small black hoppers in the 1st or 2nd stage, they were large orange ones and therefore poison bait was useless. So I had to dig a big trench, and the Chief turned out over 140 people to drive them in.

When I arrived there, I found the village was comprised of some round mud huts with grass roofs; the Chief's house was a number of huts joined by mud walls, in the centre of the village. A deckchair was produced, and with one of my overseers as interpreter I said 'How do you? Are you well? Are your people well? Where are the locusts?' The Chief himself said he would lead me there, and a push bike was produced from the Chief's hut; he carefully pumped up both tyres. The entire village, the elders – mammies with pickins at the back, gazed in wonder.

I could not hurt the old boy's feelings, so rode this pushbike over bush paths until eventually I fell off. Then a horse appeared all dressed up in leather and brass with a saddle. I've never ridden in anything quite like it before – an absolute armchair. This was more in my line. Eventually we came to the locusts, and I had a mighty trench dug and got a swarm pretty effectively. I then gave the Chief a ride back to the village in my car; he was delighted and made me a dash of a calabash full of eggs – 31…."

I then set off in the other direction to Sameni, about 15 miles to the west of Gambaga, thence to Du and Bulbia; the latter was a place riddled with mosquitoes and sand flies where I developed an attack of what I diagnosed as malaria fever, the first of my tour. This distressed me as I had hoped to be one of the few who were able to say at the end of their first tour that they had not been attacked by malaria. I had, of course, taken my usual daily five grains of quinine as a prophylactic; woe betide you if you had not done so regularly as there was a danger of breaking down to blackwater fever. I increased my dosage for a few days. My temperature came down, and although I was left feeling fairly weak and depressed (a usual symptom of West Coast malaria), I was myself again.

The rest house at Bulbia was beside the market, and pandemonium was going on over the wall when suddenly I became aware of a strange quiet. I was told by my steward boy that everyone was listening to the sound from my gramophone, which had never been heard before. There was considerable argument as to how many people were inside the box to make such a noise! The market, held every third day, was interesting; we used the market day when we wished to send word ahead to a Chief that we required a rest house. Chiefs and headmen rode on horses and the police had a mounted troop, which went on village patrol as well as being used for ceremonial purposes. There was always somebody in the market from the next village who would convey a message. Markets were useful places in which to check prices and the availability of foodstuffs, the laws of supply and demand operating very positively. Rain would cause an immediate rise in the price of grain.

When crossing the main road at Walwale I heard that two white men were on their way south from Pwalagu, one of whom was very sick and was being carried in a hammock. (28) So I waited for them. It turned out to be Jim Syme, the District Commissioner of Bawku, who had been nursed through blackwater fever by John Muir of my department. The patient was much improved and came in on a bicycle.

I finished my tour in the western part of Gambaga district, by which time the dry season was beginning to make itself felt. The days were crisper, the nights clear and the rain clouds seemed to have passed. This had the great advantage that I was able to sleep outside on my camp bed, with only a mosquito net between me and the stars. It was lovely after the heat and sour smell of the rest house which, being open and built of mud, had absorbed the heat of the day and provided a playground for the ubiquitous village goats.

Whilst in Gambaga the District Commissioner and I opened the tennis season, with a make-do net made in the prison. The District Commissioner apologised for the inadequate marking of the tennis court. I did not pay much attention to this at the time. The following day when I was looking in the prison store, the District Commissioner pointed out the drum of the whitening which had shown up so ineffectively. He added that it had not whitened the walls of the prison very well either. I remembered the missing drum of sodium arsenate, but to my relief no one was reported dead.

I returned to my work at the Research station in Tamale, running the experiments, and gaining more departmental knowledge.

29th July 19.30 Tamale "Dear mother,
As I write I hear the intermittent drumming as some chief, probably from Savelugu (29) comes into Tamale. About 20 women are trotting with all sorts of loads on their heads. There are two horses with elders on, stacks of trappings and bells and then the Chief with a big umbrella. Behind him a few more horses and then men with these funny drums tucked under their arm and a curved drum stick. These latter followers walking have probably come in 15 miles. Golly, I would not be a woman in this country, or a native horse, for anything! Much pomp attends this petty procession...."

My next trek looking for locusts took me into the mandated territory of Togoland on the eastern frontier, which was fairly densely populated with virile people, the Konkomba. I stayed in rest houses at Tundi, Tambakuragu, Bunkpurugu and Pasenkwia. There were magnificent leguminous trees in flower at the time (end of October), pterocarpus erinaceus, which stood 30 – 50 feet high and were covered with a mass of highly scented golden pea shaped flowers very attractive to bees.

10th September, 1930 "Dear mother,

We have nearly come to the end of the rainy season, the end of this month sees the finish of the heavy rains and we have 6--8" to come. The water table is still rising however, and evaporation is very great giving a delightfully misty Englishy effect in the early morning.

Unfortunately, a necessary concomitant of the rains is an abundance of arthropods – mosquitoes, sand flies, moths and flying ants which wander around carelessly over the table and fall into your soup at dinner. Bats fly around the ceiling eating wandering insects, and it is good watching a little friendly lizard catching 'abinchi'. A praying mantis just hit me in the ear. Apart from the discomfort these damn things cause in the house, I have to have a gang of a dozen boys wandering round the cotton plots crushing the larvae of the cotton leaf roller. The Economic Botanist has come down from locust work in the North, and taken up residence here in readiness for the cotton season. Some of the earlier plantings on the lighter soils are already flowering. Much data on flowering and boll records, bud and boll shed number, classification and a stamen index taken.

I got 840 lbs to the acre from a maize plot in the Rotation area which represents the highest recorded yield on the Tamale station since 1912. It is still very low however – last year's highest was 475 lbs. We are getting a crop of groundnuts from this plot as well this season, so we must not grumble.

Alas, the wind indicator, which we the Agriculture and Veterinary Departments in Tamale created, showed vertical yesterday. We had the doubtful pleasure of informing the Chief Commissioner that a swarm of adult locusts took four hrs to go over Navrongo last night; they were travelling in a southerly direction. This will be extraordinarily serious if they come much further south. They will get our guinea corn and millet and there will probably be a famine. Last year they did not come down till November when everything was harvested. I have made all the labourers bring a head load of grass every morning. I have arranged for three watchmen from 3 pm when labour finishes till dark. We have got 10 acres of guinea corn in bulk, besides our rotation plots. Provided we have sufficient warning we should be able to keep the adult swarms from settling with bonfires and beating of old tins. If we can do so this it will probably be harvested by the time any more flying hoppers can hatch out from any eggs this first swarm may have laid."

2nd October 1930 (3.20 am) "Dear Mother,

This comic hour might lead you to suspect that I had been doing a round of dinner, theatre, cabaret, supper and nightclub. But don't be misled! I had six perfectly good hours sleep, although I have been up an hour and am now sitting at the rest house table writing by the light of a single hurricane lamp. I am awaiting the dawn when I shall mount my pushbike and pursue my cook and breakfast.

My loads have first gone off on the heads of 26 carriers, but more correctly 24. I had to get eight from the Chief of Walwale. One ran and another went sick, so there are two loads left. I expressed my displeasure to the Walwale Chief, a poor little old man and a weak Chief, and I can hear him gargling and spitting now; he refuses to leave the compound until his Treasurer and Prime Minister, whom he has sent to town for two men, have returned.

I was as I told you, suddenly seconded for locust duty and thought peeved at the time, am thoroughly enjoying trekking. The Beetle took me to Nabogo Ferry last Sunday, and my loads, orderlies, overseer and boys went off on a lorry to join with the 25 carriers waiting at Savelugu. Atchulo waited for me there, with a pushbike so that I was able to breakfast with Symond in comfort. By the time I reached Diari, some 11 miles past the ferry, my loads were nearly there. I had one spot of water to cross, but a canoe was forthcoming, so all was O.K. – It was dashed funny to be canoeing over the road! On Monday I had a short trek of eight miles to Pigu, and on Tuesday Pigu to Nasia 14 miles with a long canoeing journey over the Nasia River. This is a mere trickle in the dry season. At Nasia, the Chief sent up some fresh fish in addition to the usual fowl, eggs and yams – it was a pleasant change.

Yesterday, 14 miles to this good spot! This was the rest house I came to at Easter time. I am pushing straight on to Gambaga 32 miles from here, there being only one rest house between, at Kombenarba about 18 miles away. It is a long trek for the carriers, with all my heavy loads. The District Commissioner, who met me here, tells me there is a swarm of hoppers there so I will swat them tonight. I wish it would get light, as I am being bitten to hell by mosquitoes.

The Police orderlies are very useful and make excellent batmen. My shoes and saddlery are kept spotless, and all my bush shirts and shorts they wash. I was in great luck yesterday when I arrived here. A sheep had just been killed, so I had an entire meal of kidneys on toast, followed by leg of mutton. It is a change after the eternal fowl.

I stop at every village as we pass through, and give instructions if locusts should come. I ensure a supply of the medium for spreading the poison at each village. Little Bill is enjoying the trek as much as anybody, he gallops along behind my pushbike all the way, and occasionally dashes to chase a lizard or bird.

A steely grey light is appearing in the east so 'before small' I will be able to push on. I have Atabolo, an overseer from Sule with me, and he waffles in Hausa as we go along. Fortunately, my cook has a bicycle as well, and he can get on ahead and get the rest house ready or a fire going...."

7th October 1930 Saleleu "Mother dear,

I had breakfast by the wayside, my cook making a fire in the middle of the road. The carriers took a rest, and my table and camp chair were put up and the 'Observer' to read. It was a lovely morning, and as I toyed with my porridge and scrambled eggs, it amused me to think of the number of people at home who must have been eating a similar breakfast. They would have a more recent newspaper – if we could suddenly change places what a magic carpet it would be.

I spent two nights at Kombenarba, and was responsible for the destruction of a small swarm of hoppers with poison bait giving 100% success. There also was a large swarm spread over five acres of which we got about 60%, I estimate, with a terrific trench.

I arrived at Gambaga on Saturday; it has water running all the year round. It is over 1000 ft high set on the side of the scarp, and in reality is a health resort. I rode out to Bulkurie yesterday where locusts were reported, and found that the natives had actually dug a trench themselves and killed the hoppers. I had a lovely view from the top of the escarpment over the flat country towards the Northern Boundary, into French Territory. (30) It was one of the best views I have ever seen. This morning I trekked along the bottom of the scarp to Nalerigu, four miles from Gambaga, where hoppers had been reported and I found them destroyed. I looked at the farms, and it was then about 8 o'clock.

My cook, who had gone ahead, had breakfast ready for me in the rest house. (31) The Chief of Nalerigu is the Na of Mamprusi, – a very important stool. He came out to meet me on horseback with his drummers, water carriers and kola, and a boy carrying a big leather pouffe, another with a worked brass stick. He wore a red- hooded cloak looking very like Father Christmas – all worked in gold and silver, but the worse for wear. I told him the usual hype, and how I had heard about him and looked forward to the time I should meet him. Is he well? His people? Stock and crops? Has he seen any locusts? What instructions has he had? And what has he done? It is his food the locusts will eat, and not mine, and so on and so forth.

I then had to walk 11 miles to this place Saleleu, where there is a very excellent rest house – a large round house built by Mackay last year. I have shamed the Chief into action. There are nine swarms here and I want to return to Gambaga tomorrow, then all hoppers this side of Gambaga will be cleared up. I am then going to trek for two days, through the bush to Weluga, and then on to Du and Basemkeire, then I will have been to every reported swarm. My work is complicated here by the limited use of poison, owing to the cattle route over the scarp coming through here, and the trenching method takes too long.

Heard lions roaring in the distance last night – it put the breeze up Bill who sensed it first – some instinct I suppose…."

9th October 1930 Gambaga "Dear Mother,

How the days slip by – I trekked back this morning, having poisoned three swarms at Saleleu yesterday. Got a long trench dug, and 145 people turned out to drive them in with branches. Even with two orderlies, they took a good deal of organising. I sat under a tree and directed operations from my camp chair. The trench was absolutely black with the blighters, and they extended over an area of about four acres. We also drove out two harmless snakes and a rat. I was pretty tired when I got in this morning, but after lunching with Mackay I feel revived. There is one swarm four miles south of here, and that is all the reported sightings from this side of Gambaga cleared up.

It is a perfect evening, – very still, a gentle cool breeze blowing as I sit on the veranda of the rest house, and gaze at a row of wooded hills in the distance. It is a very happy and independent life, and delightfully quiet and natural….."

19th October 1930 (Sunday) Bilibia "Dear Folks,

How time does fly. It is a week since I wrote the above; it seems like the day before yesterday. I have had some pretty stiff trekking since I am getting near the Volta here and the road is swampy. Trudging through miles of mud and water in the heat of a West African morning is a tiring occupation. Then at the end of it your rest house is inundated with persistent mosquitoes and sand flies. In spite of the frequent administration of flit they bite all day, Du was terrible last night, and this place seems almost as bad. Though a good breeze at the moment is clearing everything it will only mean a storm, and then I will have to sleep within the hut which has absorbed all the heat rays of the day instead of under the stars as I have been doing lately.

I just had about a dozen ladies up here with yaws, a very unpleasant disease, wanting injections. They heard a 'baturie' was coming and thought it might have been a doctor – nothing so active, merely a Garden Master who makes them dig big trenches and kill locusts, when Allah has sent them and presumably meant them to have them! I am known to the locals of Tamale and around as the "Long Garden Master", being 6 ft 1" I am much taller than all of them !

I wish I could have been with you today and just had an ordinary decent Sunday, – but I suppose I can't have things all ways….."

Touring was a great tonic providing regular exercise. By walking or riding 10 miles or so between rest houses, there was no finer way to see the country and to meet the people. They were without exception hospitable and friendly. The Chief often met a visitor on his boundary and provided an escort to the rest house, where one learned the village news, including the incidence of locusts and the state of food supplies.

The District Commissioner sent vegetables and fruit beautifully wrapped in a banana leaf from his garden in Gambaga, together with mail. This came by post- to- post carriers and 40 or 50 miles would easily be covered in a day. The fruit was mostly limes, papaws, guavas and cape gooseberries, whilst the vegetables consisted of lettuces, tomatoes, carrots, cucumbers and spinach. Sweet potato leaves, common in most villages, made good spinach. Tomatoes, onions and ginger were often available in the markets together with a small tuber, Fura Fura potato (Coleus dysentericus).

At Bunkpurugu I was lent a horse by the Chief, a keen well-built pony which I enjoyed riding up and down the hills in that part of the world. I was intrigued to find the people preparing fish nets for the traditional fish hunts at the beginning of the dry season. (32) The fish were caught in the lagoons as the rivers dried up, and ceremonies were accompanied by ritual whilst dried fish provided an important stock of protein.

2nd November 1930 Bongo Daa, Mamprusi "Mother Dear,

The rains have just about stopped, and crops are ripening rapidly. I am walking now on trek, since we have so many little hills. Many sandy patches make it difficult, so that little advantage can be had from pushbiking, it is a poor sport anyway over bush paths, (33) and walking is much easier, So my wretched orderly has to take the pusher. But what a good life it is and how expansive Nature is just here, where there are only two white men for miles and miles; that is Mackay the DC, and myself.

There are large areas of bush here, many outcrops of rocks and gravel. The bush consists of trees about 10 – 20 ft high, and a few higher, and grass anything up to 8 ft. The latter is getting very brown and in a month all will be black with bush fires. Villages are very scattered, about 10 miles apart, where a pocket of good soil occurs. It is this last fact that makes locust control so difficult. The Native says 'why should I go five miles into the bush and kill hoppers, they are not damaging my crops?'. And who are we – who know that flying adults hatching from hoppers now, will strip the second maize crop in Northern Ashanti hundreds of miles away – to blame him…."

5th November 1930 on trek Tundi, Mamprusi "Dear mother,

When these people are given a 'takarda' – letter – they split a stick and fix in the letter. The messenger runs to the next village and just gives it to the Chief, saying the name of the place it has to go to. Another man runs on to the next village. It is amazingly quick. I have had letters from the DC when he has been three or even four days away. That is anything from 40 – 80 miles, and written the same day. My vegetables are sent in the same way from Gambaga, and they never miss.

There is no proper rest house here, only a camp – little round grass houses, with a beaten mud floor. They are really rather pleasant, just in a clearing in the bush. The weather is perfect. I think the rains have finished. I sleep out now under a lovely moon and stars, and awake with the dawn. It is quite cold in the mornings up till 7 am, and my cold bath takes my breath away. It stokes up like blazers from 2 pm–4 pm.

I spent my time at Bongo Daa scattering poison bait, and measuring farms – I am collecting some very interesting and useful facts concerning farming practices, and trekked on here yesterday. I am now in the Mandated Area.

Big show yesterday when I arrived – I had the Chiefs of Tundi, Tumbukragu, Bunkpurugu and Bende awaiting me with their retinues, and five police constables. I delivered a long lecture on how to proceed with locust control till I have time to get to them. So I sent them off with a measuring stick and Constable to do it. Meanwhile, I am trekking round with poison, which is wonderful stuff, as fast as I can…."

13th November 1930 Gambaga "Hello troops,

Here we are back again at H.Q. and here to remain until I get definite orders from Cottrell, Entomologist, Locust Campaign. Did I tell you about poor old Cook – an entomologist who was in Trinidad with me, and stayed with me on his way up through Tamale? He has a very springy walk (called Spring Heeled Jack here). He died in pain at Wa, in the North on locust work and it is suspected from arsenic poisoning. I am awfully sorry for his people. You see, we were just dished out with a couple of hundredweight of the stuff, and our common sense. I enquired a little time ago if the lethal dose for a human was known, in an effort to calculate how many leaves of a cowpea grown by the native on his guinea corn farm would affect their health if used in soup making. I wanted to know how much he would have to eat with the concentration I use, and merely got a reply "no knowledge in this office". Fortunately, I have not killed a Mamprusi, a cow or a goat that I know of – yet – and my work here must soon be finished.

The harmattan season (34) is now on. The hot dry wind comes down from the desert. My nose is tingling and later my lips will peel. Tables and chairs, picture frames and boots all crinkle up and saddlery has to be well oiled. A haze hangs over the hills and the mornings are really cold, the afternoon really hot…."

20th November 1930 Walwale Rest House (11 am) "Dear mother,

I enjoyed the trek with Cowan. On the way from Gambaga to Nasia we saw the most grotesque show I have ever seen – a funeral custom. (35) Four figures swathed in black and red sisal fibres danced and danced as in a stupor, to the drumming of native drums around a swish mound. It had been whitewashed, and was covered with fresh blood – probably goats or chickens and likewise some heads of antelope– juju heads. It was all very eerie. These four figures had knives in their hands and looked particularly bloodthirsty. No women are allowed to see the ceremonies, and once or twice they hid themselves and the drumming stopped.

At Nasia, Cowan and I braved the crocodiles and swam in the dirty river – we took care to have an orderly with a rifle, on the bank. There would have been hell to pay if he had shot one, because the Chief informed us that all in the pool we bathed in were juju crocodiles. (36) A lovely family of little black monkeys came out of the bush by the rest house.

Since I was in Walwale last, a new Chief has been installed, as a result of the humbug I had with carriers here on the way to Du. Mackay brought the Na and de-stooled the Chief, and now I find myself showered with chickens, guinea fowl, pigeons, yams, eggs and groundnuts, and I don't anticipate any palaver again. I have sent the Chief along a road and to tell me what he sees on his return. He says he has no hoppers, and there is a swarm a hundred yards from the first compound in his town.

Big market today and I must go down and see it. Already the road is thronging with Moshi and Hausa men toting crates of chickens down the road to Ashanti, from way up in French country. (37) They go down in their thousands every year and buy kola nuts. They will call in at Walwale market; so many types are to be seen some with long spears, others with swords – some with helmets and others with turbans.

A harmattan haze hangs over everything and I am several hundred feet higher than at Nasia. There will probably not be so many mosquitoes and sand flies. They were awful at Nasia…."

The danger to crops from locusts was now over for the year, the roads were open, the dry season drifts were built over the ferry crossings and motor traffic was free to move around. I returned to Tamale feeling that I had learnt much about the country and the interesting Mamprusi.

Back in Tamale I helped John Symond with the installation of oil extraction machinery in the cotton ginnery. We were looking into the commercial possibilities of shea butter from the walnut-like fruit of the indigenous shea tree (Butyrospermum Parkii). (38) The machinery consisted of a hammer mill, steam kettle, hydraulic press and filter press, and 15 tons of nuts had been bought in local villages for the first trial. Before the conversion it took anything up to two weeks and much anguish with 20 men pulling on a rope wrapped round the flywheel to produce an inflammable gas. Unfortunately, the value of shea butter on the world market proved too low to meet the costs; also the fat was found to have too high a content of unsaponifiable matter to be attractive to the food industry. Work in the ginnery, the harvesting of crops on the experiment station and the statistical analysis of the results kept me occupied during the dry season. (39)

Exhilarating harmattan weather lasted until the end of January; during this time a tweed jacket was needed hacking around the farm before breakfast, and the air was intensely dry. Soon the wind changed from north to the southeast, storm clouds appeared and it became very sultry with temperatures well over 100° F in the shade. Dust devil carried leaves several hundred feet into the air, and if the bungalow or office was unfortunate enough to be in the path everything was quickly covered with dust and rubbish. There was much sheet lightning at this season and sporadic rainstorms of high intensity, up to an inch of rain falling in an hour. This doing almost as much harm through erosion as it did good by relieving tension, and starting the growth of vegetation again. It was a trying season for man and beast.

We had a two- day race meeting over Easter 1931 held on a new racetrack outside Tamale, with five races each day. Zana (grass) mat shelters provided for a tote, bar, tea tent and stand, saddling paddock and weighing-in room, with rails around the bend and down the straight. Of course, all this was possible with the enthusiasm of the officers in the regiment. Various prominent individuals and organisations generously presented cups or pieces of plate. Riding in the races was a tremendous thrill, but I was too heavy to be very successful. The joy of meeting in Tamale was that almost every one of the 40 Europeans present was either an owner or a jockey, or performed some function. We had a native race each day but the meeting was essentially an expatriate exercise.

Polo was popular in Tamale; generally four chukkas an evening on Mondays, Wednesdays and Fridays provided enormous fun and splendid exercise. Police ponies could be hired for l/- a chukka and the Inspector General of Police, Colonel Bamford, would only post to Tamale senior police officers that were keen. The Polo Club also had ponies for hire on a monthly basis at reasonable rates.

In late April, I went up to Gambaga for 10 days "local leave" to learn the language and do some shooting. I stayed in the rest house, which contained a most comfortable hip-bath like the one Bubbles used in the Pears advertisement. My friend from locust days, Fred Mackay, (40) was still the DC, and his charming wife Sonia was with him, having left their two boys at school in England.

Most mornings I went out with a hunter from the town. The bush was green and fresh at this time of the year. Grass was low and shrubs and trees were getting their new leaves, long shafts of light from the rising sun showing up the glistening dew on cobwebs. Watching the hunter was fascinating; clad in a short brown smock and hat he walked, almost ran, with light quick steps, stopping now and again to listen and test the direction of the wind with a little sand, to look at some nibbled grass or examine recent spoor. Flocks of cackling guinea fowl crossed our path from time to time making one feel the world was inhabited, but of course they warned other game of our presence. Dog-faced baboons were curious, and travelled parallel with us barking to the annoyance of the hunter who commented "biri ba da kyau" (monkeys no good).

19th April 1931 on local leave Gambaga, "Mother dear,

Coming up on the lorry my boys spotted a gorgeous bush meat, a roan antelope. I thought it was a mustard horse. He stood about 40 yards from where the lorry stopped – curious – while I got the .303 out. It then moved off. We went on about two miles and there he was again, he must have run parallel with the road.

He crossed behind us when we stopped. I got a shot at him running, and missed. Bill (the dog) ran after him, and 'chased him for bush'.

Yesterday morning I got a hunter from the town, and was up at 4.30

and off by 5; we went up over the scarp. It was a lovely morning. I got the real thrill of hunting, working upwind of two duikers feeding in a glade. Got within 40 yards on hands and knees, and then fired over them. The hunter then took me to a salt lick, a most interesting spot in the valley. The ground all around was poached with fresh tracks of countless bush meats. One rock stood up in the middle, and all around the rocks were licked smooth. I ate a sausage roll at this juncture and drank a bottle of water, and then on again. We disturbed a herd of about 20 hartebeests, they didn't actually see us. We followed these for about 300 yards from tree to tree – I missed a beauty again at 70 yards.

We saw the odd duiker, heaps of fresh spores. Lots of brown and black monkeys of all sizes who amused me immensely, springing up and catching hold of a branch of a tree, and watching us over the grass. We got back about 11 am having tramped about 15 miles. The hunter was a joy to watch, picking out tracks, broken twigs and nibbled grass. We found one place where he told me two 'keuki' – hartebeests – had fought that morning. Since he can only speak Hausa and no English, it did my language good.

The 'Tendana' (41) or headman of Gambaga has by custom, the right hind leg. He had sacrificed a fowl to the town juju, to bring me success. The hunter has the right shoulder, the Sergeant of the Constabulary, the school master – in fact all the local celebrities – get their dash of 'nama', Hausa for meat. I am off tomorrow over the scarp to the River Volta…."

Back in Tamale, one became aware that the economic depression was making itself felt in the Gold Coast. The territorial budget showed a deficit of half a million pounds, out of a total budget of five and a half million pounds. All departmental estimates were cut by 10% across the board. Customs duties which were the main source of revenue were increased to 20% ad valorem. There was much talk of retrenchment. Two specialist contemporaries of mine in Trinidad, engaged on a salary of £600 per annum, were made redundant and re-engaged as Inspectors of Plants and Produce on temporary appointments at £480 per annum. They were subsequently reinstated but it was all very unsettling at the time. All new works were stopped, and the Public Works Department lost 50 engineers and architects. Early retirement was encouraged by favourable abolition terms. Efficiency was judged by the negative standard of the amount of money not spent. (42)

On the King's Birthday, 3rd June 1931, the Regiment performed the ceremonial parade and the Chief Commissioner, Major F.W.

Jackson, took the salute. Most of the troops came from the Northern Territories, and extremely smart they were. The Chief Commissioner's car was accompanied by a mounted police patrol carrying lances and pennants. There is something about a military parade of this sort that stirs the finest and deepest feelings of loyalty. We played a Civilian v. Military Polo Match in the evening.

So I came to the end of my first tour.I greatly enjoyed my journey home on Elder Dempsters's ship R.M.S. Abinsa; I left Takoradi on 3rd July embarking at Liverpool. The gentle journey home was a great tonic; leave - a week for each month of the tour - started when one set foot in the United Kingdom and the Crown Agents booked one's return on the first boat after leave ended. Leave was generous but absolutely necessary for a complete recharging of the physical and mental resources.

I had news of Tamale from my colleague Gordon Cowan that it had become a sad place during the rains with the deaths of five Europeans out of a population of 30. (43) This was as a result of yellow fever, the virus carried by a Stegomia mosquito which breeds near habitation. When one had known the individuals concerned personally the news had a greater impact. I had handed the Secretaryship of the polo club to Geoffrey Thompson, in the Education Department, and he was one of the first to go. Another was the charming wife of Long John Miller, the District Commissioner. I went to Burroughs & Welcome in London for protective inoculation; they first took a blood sample and found that I was already protected against it. I wondered if what I thought was malaria in my last tour had in fact been a mild attack of yellow fever. This was a sharp reminder that West Africa was no health resort.

Chapter 3 - 2nd Tour 1931–33

Move to Zuarungu, start of survey (44) and Ejura groundnut scheme

Leaving Liverpool on the 3rd December 1931 I returned to Accra on the Appam. This time we had a quiet uneventful journey. I learnt that on this tour I was to be sent further north still to Zuarungu, in Fra-fra country, (45) with broad terms of reference to investigate the alleged shortage of food. (46) In previous years a number of Fra-fra families had been moved from the densely populated North to the relatively empty country around Salaga, 150 miles south. (47) Not surprisingly, the people ran home as soon as the Escort Police were withdrawn. It was not appreciated at the time that they had been moved from granite and greenstone to sandstone, from grain country to root country.

A visiting dietician from Nigeria, Dr McCulloch, advised the Chief Commissioner to send up an agricultural officer to investigate. Nobody in the Department had been stationed north before, except during the previous year on locust control, so I had the opportunity to break new ground. An early lesson was learnt that one cannot successfully force people to move their habitation, only persuade them when the course is proved right.

Before going north I stayed in Accra, to play polo for the Northern Territories in the annual match for the Tamale Cup. I dined with the Governor, Sir Ransford Slater, and Lady Slater (48) at Christiansborg Castle with the surf beating on the rocks below. It was a wonderful experience for a junior officer about to be posted to a remote and interesting assignment. The Governor's ADC, Blackburn-Kane - known to the Africans as "Black Monkey" - was suffering, poor chap, from a self-inflicted broken ankle. It happened while taking a running kick at a cannon ball mounted on the ramparts.

30th January 1932 Tamale "Dear Mother,

We had a great night at the Castle before we left Accra. His Excellency and Lady Slater are most humane people, and make one feel thoroughly at home. We had the Police Band again playing 'the Roast Beef of old England' as we went in, and they played all through dinner. His Excellency proposed a toast to Polo, after the King. After dinner we played progressive ping pong. Lady Slater made us remove our dinner jackets. Snode then performed on the piano, and after we went out onto the bastions, a lovely moonlit night, and sang all sorts of songs with the sea crashing away at the foot of the castle below.

I said the wrong thing to His Excellency when I saw a funny little dog sniffing round when we arrived – I enquired whether it was a stray, or belonged there. He immediately turned to Lady Slater, highly amused, and said 'Lynn has suggested that your dog is a stray.' I heard afterwards that the dog is the joke of Accra.

Things are pretty grim here. No travelling allowances, threats at a levy on salary and retrenchments, a percentage of people going from P.W.D. including Lloyd. Where he will get another job I don't know. Three of our Department, Eady, and Parkes being retired on abolition terms and Bowen, a new man who is in Accra hospital with phlebitis poor chap...."

15th February 1932 Tamale "Dear Mother,

On Thursday, we went up to Navrongo to see Percy Whittall, the Provincial Commissioner for the Northern Province. (49) Percy is a great character; he has been up here for a long time and is just retiring in May. He is a great potter and has made a brick works making tiles, pipes, rough jugs and things. He is very proud of a new hospital he is building there with his bricks and tiles. Of course, he has stacks of free labour, and about 20 bush masons. The Governor will open it in March when he comes up.

I was very impressed by the White Fathers Mission at Navrongo. (50) They have a very fine swish church and school, and spend all their lives out here. Father Morin had been out 18 years before he first went home. They are Roman Catholic of course, from Canada I believe. They really know the native and force nothing. I went into the church and little kids would come in with just a loincloth, cross themselves, and pray apparently in all sincerity.

On Saturday, we came 18 miles back along the Tamale road, and then branched off to Zuarungu where there is a District Officer and a doctor. The doctor is moving into Navrongo when the hospital is finished; here is where I propose to make my headquarters. The country is more open than around Tamale even, and every square inch practically farmed.

Behind Zuarungu there are the Tong Hills, where there is a great fetish cave. People come from all over the country, sometimes hundreds of miles to make a sacrifice. It is usually a fowl or goat and they ask for some blessing. On the other side in the distance, to the north, are the hills in French country. The country is gently undulating.

The people, Fra-fra, (51) are quite different to either the Dagombas around Tamale, or the Mamprusi in the Gambaga district. Instead of making their houses grouped together in a village and making them with grass roofs, they make compounds of a few flat- roofed houses. These compounds are scattered one one all over the landscape, usually 10 to 200 yards apart. The women only wear a bunch of leaves, or beads, whilst the men dress themselves up. They almost invariably carry bows and arrows, with which they are very skilful. They have an extensive knowledge of plant poisons as well...."

In Zuarungu, where the population was most dense, a house was becoming available. The main effort would be concentrated in that district, with preliminary observations in Navrongo. The work was eventually to carry out a comprehensive agricultural survey. This was to include systems of farming, areas of land cultivated, varieties of crops grown, methods of storage and disposal, manpower available, marketing data, wild sources of food, livestock in the economy, water supplies, geology, soil and weather conditions, followed by field trials with possible improvements. (52) A detailed programme of work was submitted for the Director's approval. The Chief Commissioner strongly approved the programme, the survey being in line with the policy of indirect rule through Native Administration. (53) This was receiving much attention at this time in place of direct rule, and factual information was badly needed.

As I prepared to leave Tamale the Public Works Department was about to turn on the running water there, pumped from a dam between our ridge and the town. A 3 ft crocodile caused a delay by a burst in the mains. Electricity was talked of for the near future, and new houses were wired accordingly, altogether very sophisticated. I had a personal sadness before leaving in that my steward boy, Jacob, died of tuberculosis in Tamale Hospital.I left the comfort of Tamale and travelled 98 miles north to my new assignment. I was familiar with the road up to Walwale from my locust day's 18 months before.

Leaving Gambaga District, down a steep escarpment at Pwalagu, one sees the Zuarungu District with its scattered compounds over a vast plain lying beyond the Volta River. The effect was quite spectacular and marked the change from the Voltaian sandstone to the south, and granite and greenstone to the north. The river crossing at Pwalagu, (54) 18 miles from Zuarungu was by a dry season drift. This was a sort of Irish bridge which was washed away by the rains in May, and built up again on the old stone foundations in November. During the rains no vehicle could pass, and mail and stores were ferried across by dugout canoe. Only very persistent visitors came north during the rains.

3rd March 1932 Tamale "Mother dear,

I was supposed to go north to Zuarungu next Wednesday but yesterday the news came through that Pwalagu drift had gone at a point 30 miles from Navrongo, as a result of a tornado and 3 inches of rain. It is the first time anyone can remember the drift going earlier than May. It has since been repaired however, and I have decided to push up at once. So today has been an awful scramble, not so much for me of course, but commotion in the house. The end is in sight.

The anticipated 'levy' comes into force next month. Under £600 married 5%, unmarried 6%. Over £850 married 7% and unmarried 8%. Over £850 10%. Meanwhile Conservancy Fees have been raised 200%, ad valorem import duties to 20%. (I had £16 additional to pay – stiff). All this is because a spineless and corrupt Legislative Council are afraid to put on an income tax. This would have affected all alike Native, Syrian and European. There was a demonstration at Sekondi, Cape Coast, where the Provincial Commissioner had a stone put through the windscreen of his car. The tax was taken back as a result. All travelling allowances have been stopped, as well as seniority allowance…."

The Government station at Zuarungu (55) covered about 100 acres. The buildings were built with sun- dried bricks, roofed with thatch and comprised the District Commissioner's office and court, a prison built like a Foreign Legion fort. There was a small police station with the police lines behind, also providing accommodation for 12 constables and a sergeant in large roundhouses, and a smallish hospital. From the courthouse an avenue of flamboyants (Poiciana regia) led to the District Commissioner's house half a mile away. It has a Union Jack flying from the flagpole. Half way along the flamboyant avenue the road forked to the two rest houses and to the doctor's house. This was to become the Agricultural Officer's house when the doctor moved to Navrongo. The flamboyants were in flower when I arrived, and the sight of a bright blue roller bird in the trees against the deep blue of the sky left a lasting impression. The view to the south from the houses was to the Tong Hills, six miles away and very impressive. To reduce maintenance of the Government land the District Commissioner allowed the local people to plant groundnuts, which left the surface very bare and stony for most of the year. The water supply for the station was a crocodile- infested waterhole in the valley, where the vegetable garden was located.

I took up residence in the three large thatched mud-walled huts, (56) which constituted the rest house. The first was the bedroom and bathroom, the second the dining room and pantry and the third the lounge and office. The kitchen and servants' quarters were separate. The floors were a warm red colour made from beaten earth mixed with rotted cow dung, and the red infusion of the pods of the dawa-dawa tree (Parkia filicoidea), a carefully preserved component of the local flora. The local women were skilled in laying floors and plastering walls. The inside walls were a cream colour obtained from locally mined kaolin, using the mucilage from okras, a local vegetable, as size. There were openings for doors and windows, so the entire construction was built without nails of local materials by local artisans and labour. With camp furniture and a few native cloths and mats, it was surprising how cool and comfortable a simple roundhouse could be made. Beside an open air stoep grew a strophanthus bush used by the locals as an ingredient for the poison on their hunting arrows.

16th March 1932 Zuarungu "Mother dear,

Here I am well settled in Zuarungu, with my programme gradually taking practical shape, and the body very fit.

The natives are waiting for rain before planting their early millet, so I am being given a respite to get my staff in working order.

Today, I gave myself a change going up to Bawku, the most northeastern station about 40 miles along the road from here, with Dr Saunders, the Medical Officer here. (57) Syme is District Officer there and he took a toss from his horse last Saturday and had slight concussion, and was unconscious for five hours. He was up today and gave us lunch, but is of course still on the sick list. He is a jolly nice fellow, and has invited me

51

to blow in on him at any time and stay with him, supplying me with a bed etc.

Holman the Vet is at Kunaba, west of Navrongo and I am going to stay with him next weekend and hope to get some shooting on Sunday morning. He is running a rinderpest serum station there, and his cattle patrols say there are lions five miles from the camp.

I have been tracking down a lot of trees lately, with comparative success. I am rapidly gaining knowledge of the local flora which is immensely valuable, since so intensively cultivated in this district with practically no bush areas. (Very different to Southern Mamprusi.) So wood for buildings, implements, fuel etc. is scarce, and the trees which are allowed to remain are almost entirely of economic importance, either as food, flavourings for soup, strings or medicine.

The natives amuse me too; in their villages they have a little ring of flat- roomed, round mud houses, built like little fortresses, which indeed they are. They have flat roofs, not for lack of grass, because in a place like Zuarungu many compounds have the usual grass roof. It is so they can get up on them and get a good view of the countryside and discharge their bows and arrows, being protected from without by a parapet. Also, arrows dipped in oil and fired in the air, lighted, cannot drop on roofs and burn them. They even have a wall in front of the door of each house for this purpose and anyone entering creeps round it. (58)

Talking of hospitals reminds me of Saunders, who is, I believe, quite a good surgeon. He dislikes Zuarungu, where he gets no interesting cases. The only thing here is yaws. Anything else he says he has not the equipment to treat, even if he had the necessary staff of dispensers and nurses. He threatens to clear the Northern Province and send them all down to Tamale Hospital, and put himself on the sick list. I have told him that I consider he is mad, and that when we are the last two up here, I will take him down in a straight waistcoat! He is an amusing bloke nevertheless.

Affairs of the colony don't improve, and a 5% levy is to come off April salaries onwards, – this with no allowances, and an income tax when the Legislative Council has the guts enough to put it through, or turn out the regiment – makes life very difficult and hardly worth serving all one's life under these conditions...."

I arrived at a very suitable time just before the rains. The country was hot, dry and bare, with only the whitethorn (Acacia albida) showing green leaves. One wondered what the livestock could find to eat at this season, as they were all in poor condition. An important quality of local livestock, and indeed of the people themselves, was the ability to adjust to levels of nutrition and to pick up quickly in good times.

The countryside was much more open and undulating than in Dagomba and Mamprusi, and the people were quite different. Instead of living in towns and villages and going into the bush to clear farms, the Fra-fra (consisting of three tribes, Nankani, Tallensi and Nabdams) lived in scattered compounds at least an arrow shot one from another. They farmed intensively and continuously the land between. The Dagomba and Mamprusi were root eaters whilst the Fra-fra were grain eaters, mainly bulrush millet, and sorghum. The basic difference was geological, Voltaian sandstone and laterite found in the South and granite and greenstone, with quartz intrusion, in the North. The latter gave harsh, light and stony soil, with rocky outcrops and clay in the valleys. Water supplies were much better in the North, the water table was at 20 - 30 feet in the dry season, whereas there was virtually no water table in the sandstone and the people relied upon surface water in rivers and swampy areas.

The life of the Fra-fra was self-sufficient and highly integrated. All families farmed, built and repaired their own houses, often helping one another with food and the local beer, pito. Individuals developed skills in crafts such as potting i.e. making water pots, basins and household utensils. Iron smelting made hoes, spears and arrow heads. Leather work, cotton spinning, weaving and dyeing, basket work and rope making were other crafts. There was not much difference between the way of life of a Chief, headman or a commoner except in the scale of living. Consistency within tribal custom was essential, no deviation being tolerated. All this had an important bearing later on, when it was desired to bring about certain changes through extension work.

Another feature of the northern people was a lack of clothes; (59) the men wore a simple woven loincloth or sometimes a goat skin and a short embroidered cotton smock with ample trousers like jodhpurs and a loose tamoshanter type of cap on special occasions. Chiefs and headmen wore flowing Mohammedan robes. Women wore a thin leather girdle with an ornate leather tail at the back and a bunch of leaves or a finely woven web of strings in front. On special occasions, perhaps going in to the market, they would drape themselves in brightly coloured imported cotton cloths with elaborate designs. A common sight in the morning was women going out to collect wild herbs for soup with a beautifully made basket on their heads, smoking a large clay pipe with a wooden stem a yard long, tails and leaves swinging like kilts, laughing and shouting greetings to one another.

They were certainly a happy, cheerful people and the women particularly were very hard working. They prepared the food for the household daily from the materials supplied by the head of the family augmented by the wild foods gathered, often laboriously, from the trees and herbs around. Grain was ground on a stone to coarse flour; various dried leaves were stored in pots for soup ingredients. Groundnuts were pounded in a mortar and pestle to make oil and sweetmeats of various kinds. Cowpeas and Bambarra beans (Voandzeia) were used extensively in soups and stews. Salts known as "kanwa" were extracted from the ash of early maturing millet stalks, and certain swamp grasses, by a process of filtration and evaporation using old clay water pots for this purpose. These mineral salts were also used for seasoning soups.

Fowls, guinea fowls or the flesh of goats or sheep were the basis of stews and sauces or any other meat or fish as and when available. The raising of fowls was an important occupation of the Fra-fra; termites were carefully collected in old pots filled with dung inverted over a white ant run, and these were then fed to growing chicks. Drinking water was always available in pots around the compound, a common cause of mosquito breeding. Many a young man going south to work would take a load of up to 30 fowls in a long crate made from guinea corn stalks. Fowls always fetched more in the local market than guinea fowls, perhaps 4d instead of 3d because guinea fowl could not be transported in this way.

I was told by the Veterinary Officer, Stewart Simpson, that the Chief of Nangodi had informed him that his favourite meat, next to man, was dog! But the people were certainly not cannibals, seasonally fresh fish would be available and dried fish as a relish was highly esteemed.

I now started to develop my agricultural survey of the area. In consultation with the District Commissioner agricultural staff were posted at five centres; Tongo five miles south in Tallensi country, Nangodi ten miles east in Nabdam, Bongo eight miles north and Sambruno 12 miles north-west, which together with Zuarungu were Nankanni. (60)

At each centre, all the land farmed by six families was chain surveyed. Unfortunately, farms could not be taken at random because I had to select individuals prepared to work with me, and stand up to limitless questions. I worked through the Chiefs and took their advice. On closer examination, the numbers selected were reduced from 10 to six at each centre, in the knowledge that I was working with a group that was probably somewhat more efficient than the average. In the survey I established close contact with the families concerned.

13th April 1932 on trek at Bongo "Dear Mother,

The hour is 4 pm and I have just finished lunch. It is devilish hot too. Relays of small boys have been rushing backwards and forwards with bottles of filtered water, which is just exuding again from every pore in my body; my second Aertex vest is soaked through since a bath on returning.

I have been surveying farms. I wish I could think of a better way of obtaining the information I require, it is very laborious and one can't leave it to the African, for when you come to plot it, it won't work out properly at all. Churchill quotes a Spanish saying in 'My Early Days' to the effect that only Englishmen and mad dogs go out in the sun at midday, and certainly here everywhere is utterly deserted from 10 until 4 except for sleeping figures under trees. However, I am feeling very fit and interested, which is the main thing.

Well, well, here is Abugari with tea and I have my overseer bloke coming at 5 o'clock and I want to plot the farms surveyed today. There is no sanitation here in the villages at all, and the rest house is very close to a number of native compounds and full of flies. I am rapidly losing my temper with them, so I will give this a rest for a while, and try and add a little more before I send the messenger into Zuarungu with the mail...."

16th April 1932 Bongo "Mother dear,

I got in for lunch today, with a staggering thirst well after 2 pm, having surveyed a big farm. I am getting better at it now, with practice. This is where my old Wye experience is standing me in good stead. I also had two bright ideas:

1. What about an export industry in Hibiscus fibre, I must go into this. A small boy making string with it under an ebony tree suggested it to me. (61)

2. Why not construct a simple plough. These people keep bullocks merely for social reasons (30 fowls = 1 goat or sheep. 10 goats or sheep = 1 bullock and 4 bullocks = 1 wife.) But I can see no reason for not making them work. Then greater advantage could be taken of the first rains, and

55

all the land cultivated ready for sowing millet with the second. As it is they give very little time to cultivation, and complain that they have no time to hoe, so a man goes round with a flattened and sharp stick, – rather like a narrow bladed oar — and makes holes and women follow with a calabash of millet seeds and cowpea beans, and drop a few into the holes. This they cover deftly with their feet, and a shallow hoeing is given when the crop grows up as opportunity occurs. The ground is, of course, too hard to cultivate before rain. I am frightfully pleased with the idea.

I had a ripping letter from Symond yesterday; alas we may be losing him. He has been ordered to take over as Provincial Supervisor of Ashanti and to move to Kumasi on 6th June . He is furious and has got the Chief Commissioner to write to the Governor direct, and says if there is no way out, he will apply for a transfer. He was very pleased with my first monthly report, and suggests that in years to come, every Fra-fra will have a lithographed portrait of me in their compound as the 'Saviour of the Race', which of course gratified me as much as it amused me.

I felt I had accomplished something when I came into lunch today, a very satisfactory feeling. But one day is just the same as another. I felt I would have liked to have had an afternoon on the river with tea on a green lawn, or a game of cricket or a tennis party; instead I had to console myself with typing out a few pages of my old Botany notes. A little rain and much thunder about 5 o'clock have nicely cooled down the air. I am inside now. I always dine outside, and sleep. I wish it wouldn't thunder and lightning though, whenever it rains in this country. Everything is on a superlative scale. Too hot, too cold (not often, but utterly miserable without a fire sometimes). Too windy. No rain. Too much rain at one time, washing drifts and bridges away and plants out of the ground. Though it's not such a bad life I suppose really. I'm not grumbling.

Sad news. Little Bill disappeared early yesterday morning soon after I went out surveying a farm about 6 am. The messenger has been scouring the countryside and I have jerked the Chief into action, but to no avail. He may possibly have trekked back to Zuarungu on his own, that is my only hope. I have offered a 10s- reward. The trouble is that these dammed heathens eat dogs, and Bill was fine and meaty. Abugari says he would sell from 4s to 6s, and it is quite possible therefore that he has been stolen. Then there are the hyenas in the rocks, who come out at night and pounce on stray dogs, sheep and goats. Poor little feller...." (62)

It became fashionable to have this strange white man coming onto the land and asking strange questions. I soon became inundated with requests for attention, and some of the "lucky ones" would even renew the guinea corn stalks I used as survey pillars during the survey! I was greatly touched in Sambruno some years later when I stepped out of a car, and a young man came up to me with a glowing ember for my pipe. I was reminded that when I had been surveying there I had run out of matches. (63) Yield data was obtained from demarcated plots of a fortieth of an acre; again the distribution could not be at random but they were located as far as possible in representative areas on the surveyed farms, two per acre. They were numbered serially for identification according to distance from the compound. This was necessary because all activities were recorded for each plot, hoeing, planting, and manuring if any, and since all crops were mixed and matured at different times it was important to distinguish each plot.

26th April 1932 Zuarungu "Dear Folk,

Many of my seeds are germinating on the observation plots, and all the farms are looking fresh and green with the young millet 6 inches high. Many trees are putting out new foliage just now, and there are some very beautiful trees here, standing out on their own, so that their proportions can be duly appreciated. This type of country is aptly named 'Park-Savannah'. Amongst these notably are the Tamarind, Silk-cottons, Baobab and Ebony; the flamboyant or flame tree, (Poinciana regia), which is exotic, is planted in avenues in the European Residential end of Zuarungu, and is still a marvellous picture. These trees are 20 ft high with wide spreading branches, light feathery foliage and terminal racemes of scarlet or orange flowers, some 4 inches in diameter. They are really indescribable.

My knowledge of the local arboreal flora has enlarged greatly, and gives me great satisfaction and interest. Every tree or herb has its own name, which every native knows, and many have economic uses all of which I am recording. I want to produce a vegetation map of the district if possible.

The only snag to my interests is that I fear they will rather clash with types of interests and pursuits the Honorary Director of Agriculture may consider they ought to be. However, I am not worrying, neither am I letting my enthusiasm for farming, as opposed to anything else, bias or prejudice my ideas concerning the problem. When once convinced that I am right and sure of my data, (and I am taking active steps to ensure that the latter is reliable) I will submit my report, and to hell with them. I would be prepared to carry my case to the Secretary of State if necessary.

But all this ranting! Ignore it. It is merely the flow of thoughts in a mind unrelieved for some time by the point of view of another. Whilst reading is a source of food for thought it may or may not be a palliative, you can only absorb or reject, you cannot ask questions unfortunately. (64)

I fear poor little Bill has been chopped and I miss him very much. He returned to the rest house here, and a fool of a constable didn't catch him and send him back. These heathens prefer dog to all meat. I am in disgrace from Abugari too, for feeding a cat with the meat, yam and all the gravy I did not eat at dinner last night. Abugari's grumble is personal, in that I expect the remnants from my table are his perquisites; Abugari took almost joy in telling me it was the same cat who stole 10 eggs last night. He was not quite so joyful, however, when I told him to decide between himself and cook as to who should pay for them, for leaving them where the cat could take them!

Poor Abugari, it is very difficult not to become more familiar with one's servants than perhaps is good, when there is nobody else to talk to or tell of something funny that you've just read or seen, to explain why you are laughing. I tried to make Abugari say 'turnips' and 'blancmange' at lunch today. I believe he suspects me of being a trifle non compos at times. But then, all white men are mad!

I hear the rattle of a tea tray it must be four o'clock. Abugari makes a very good servant. Once told to do a thing or set of things, he goes on doing them with perfect regularity. I could go for days and not give an order. He keeps things very clean and produces a vase amassed with the Flamboyant blossom, mentioned above, every morning.

The Blanchard pressure kerosene lamp, an ornate brass affair dashed me by Dr Saunders, is a tremendous success. I am a great believer in a good light at night, it is the only thing to own, if wanting to read and write in the evening. I am altogether too self- satisfied these days.

I can well imagine myself in later years being a club bore, giving sound advice to young empire builders, 'When I was in West Africa…'. The trouble is that sound as the advice may no doubt be, it is very seldom followed and appreciated at its true worth until you have actually experienced the conditions for yourself. The typing out and revising of my old College notes are proving of great value. What is astounding me are the different values I put on different facts in my notes, to those when I wrote them down. It suggests perhaps that some improvements could be affected in the 'putting over 'of the stuff…"

In addition to surveying farms at five different centres, observation plots were established at Zuarungu, Sambruno and Navrongo so I was busy during the growing season. I used a light two - stroke Francis Barnett motorcycle to get around. It travelled well on bush paths and on motor roads in the rains, when there was no traffic.

Supervision was essential because few of my staff were experienced, and they needed constant guidance. I concentrated on the observation plots at Zuarungu where I was able to secure about 20 acres on the west side of the Government station. These plots were mainly to test the effects of compost on the local mixture of crops. There were also a number of green manures tried out as well as imported food crops such as cassava, pigeon peas and rice, which had possibilities. The aim was to obtain quantitative data in place of the generalisations of the past, many of which were fallacious and misleading. It soon became apparent that the so-called farmers of Zuarungu district were compulsive farmers, that is they farmed according to custom. Farming was only a part of their total life; they were not farming from choice or for profit. Possibly only 20% were dedicated farmers. This, of course, did not detract from what we were seeking to do; it simply meant that certain individuals were more likely to respond to improved farming methods. It was not, as was sometimes suggested, that the Department of Agriculture had favourites upon whom they concentrated.

I took over the vegetable garden from the District Commissioner. We grew reasonable crops of cabbage, tomatoes, dwarf beans, lettuce, radishes, onions, melongene, cucumbers, melon and pumpkin but were somewhat limited as regards fruit. All we could grow were coarse varieties of guavas, pawpaws, Cape gooseberries and limes. Sutton's in the UK sent out sealed tins of vegetable seeds once a quarter to each station, supplied against an agricultural vote. We could not grow the finer citrus fruits, avocados or mangoes at our low elevation and high temperatures. The garden in Zuarungu was watered from the waterhole from which the station drew its supply. The pools were infested with crocodiles, which were in some way related to Asubiri, the garden boy. One could hear them tunnelling, and from time to time they would emerge at night and lay a batch of eggs on one of the beds – rather like hard boiled hens' eggs with a soft shell.

During March and April, the weather became progressively hotter, temperatures went to well over 100 $^\circ$ F during the day, only dropping to about 80° F at night. Clouds came up from the southeast seeming to follow the course of the Red Volta River, but little rain fell until towards the end of April when, with a tremendous roar and dark sky, first a violent wind blew causing a dust storm, soon to be relieved by glorious rain falling heavily, often over an inch, in a few hours.

The following morning the whole countryside was transformed; everybody was out planting seeds, and the men used oar-like staffs with which they deftly excavated shallow holes into which women dropped seeds mixed in a calabash. Early maturing bulrush, millet mixed with cowpeas and then, in between guinea corn or late maturing millet mixed with "neri", a spreading oil seed plant. With germination cultivation was started, generally working from the compound out with a small hand hoe giving a light scarifying to kill germinating weeds. What was formerly a fairly barren- looking countryside was transformed with crops planted on almost every inch of it. The irregular farm boundaries were planted with rozelle (Hibiscus sabdariffa) which grew into an attractive red- stemmed plant; the leaves and sepals were dried and stored to be used later as a relish in soups. The bare nature of the soil, and frequent trampling during the dry season, resulted in severe wind erosion. Heavy storms resulted in severe water erosion, during and after them, until the germinating crops protected the soil – this was clearly a problem needing attention in the future.

29th April 1932 "Mother dear,

On Good Friday, I was out on trek at Tongo for a couple of days south of here where I have an overseer making records. I got Tingoli, the Chief of Tongazugu (sic) to take me up to the great Tongo Fetish with the High Priest; it was a rummy business, and talk about rock climbing! Certain aspects of the Tong Hills remind me quite of the boulder- strewn Rough Tor or Brown Willy, and there are heaps of balancing stones. Well, we climbed up the face of the hill where rough steps have been made, and came to a good- sized open cave. The Chief and elders divested themselves of their garments here, needless to say I didn't, and up and in we went.

There was not much to see beyond a large heap of feathers, old bones, cowrie shells, and a blood- splattered sacrificial stone. I saluted the fetish and wished the Chief and his people health and good crops, and came down. The Rock did not speak. (65) I dashed the Priest in order that he might make a suitable sacrifice, and the Chief then asked me if I would like to see the other one – (fetish). I did not know there were two. I agreed not to speak, silence being necessary.

I was conducted up a more arduous climb still and came to a deep cranny in the rocks, with again the heap of bones and feathers and a large number of musty garments hanging up. I enquired, when I got down, as to whom the garments belonged, and was told they were the robes of the fetish, and realised that any more questions would not help me.

The Chief then took me round his compound, a very large one up in the hills. It was very interesting. There were men of every Gold Coast tribe in there, waiting to go up and consult the fetish. King Prempeh, Omanhene of the Ashanti, made his salutations to the fetish when he returned from banishment.

My work continues up here to be both interesting, and I suspect, eventually of value. Thanks to Symond I have been given a free hand to attack the problem as I see fit, anyway at least for the first season. The Director of Agriculture sent up a comic screed saying he considers it unlikely that agricultural measures will solve the problem. It is, in his experience of the natives, not feasible to even commence this work which would be a waste of money, saying he will do nothing without legal sanction. If I can decide that 'A land settlement scheme' (66) is applicable he will put through an Ordinance; and if I consider that these measures are inapplicable then I should be recalled to Tamale. Unfortunately, all my observations so far go to show that it is essentially agricultural improvements which these people require, – a bigger variety of crops, rotation and better cultivations which a higher standard of farming implies. I say unfortunately, but actually it is the agricultural aspect which appeals to me most. From the Honorary Director of Agriculture's standpoint though, it is not spectacular enough. However, I am remaining up here because, as Symond rightly says, I could not possibly submit a report in less than a single season.

The rains are on the verge of breaking, and the result is sudden winds that swish one's papers all over the place. This cools the air for a while, and then a calm and terrific heat. However, I find that if one has plenty to do one does not notice the heat so much.

Here is Abugari with the tea tray and the afternoon gone already. I sat and read the Observer after lunch for a while. Tonight, I am going to Sambruno – six miles away where there is another overseer. I will look at the observation plots he is preparing, and he will pay his labour. Then tomorrow I will trek on to Nangodi, where more observations are being made..."

24th May 1932 Tongo Rest House (67) "Dear Mother,

I am enjoying my quiet sojourn here in spite of the flies. There are some lovely hills all round. My practice is to ride out in the early morning on my horse, where the farms are to be surveyed, and get on with it until 8 am. By this time my cook will have arrived with two carriers bearing table, chair and cook box, and prepared breakfast. He will have laid the table under the friendly shade of a 'Parkia filcoeidia'– the dawa - dawa or West African locust bean tree. (Its habitat is not unlike that of a cedar.) I am usually ready for breakfast, and surely no cook but a native could serve such a good breakfast under these conditions; porridge, eggs and bacon or sausage. This morning Edi optimistically produced four fried eggs sitting on four rissole things made of guinea fowl, flour and mustard. He said in Hausa 'because of your work, I do so'.

Ryvita and marmalade and lashings of coffee; one pipe and a story in 'Blackwood's' and then I finish the surveying, getting back at lunch time.

The afternoons I spend plotting from the field book, and when the six farms are finally finished I return to Zuarungu to complete the area. It now only remains to survey six farms at Nangodi. Then I go to put one tenth-acre plots down all over the district to supply yield data. I want to trek all through the bush areas to ascertain potential farming lands.

It is all very agricultural work and extremely interesting, so I do feel I am living. The people seem very pleased that an agricultural interest should be taken in them. I have worked on the lines of personal contact, and now my desire to really assist these people has become as much a personal matter to me, as a departmental one. The only thing that disturbs me is that I may be drawn away before I can do anything. Likewise, that somebody else will carry on the job who will not appreciate my line of work, and so belie the confidence of the people which I consider to be of paramount importance. But Departmental Administration does not, or cannot consider this to the same extent. However, the situation has not yet arisen. As Symond said when he came up in February, 'it is a wonderful opportunity to become an authority on this part of the world, and so consolidate your position in the Northern Territories.'

These people can't quite place me yet. They only know two or sometimes three categories of white men – DC's (68), doctors and cow doctors and so I get people brought to me who have stolen, appropriated his neighbour's wife or tried to commit suicide. Cases of yaws come for 'pumpy' i.e. the hypodermic, or herds of cattle appear in the compound. As the Agricultural Officer I am convinced that if we can only do some improving either by seed selection, importing of new crops or industries, then the native will not be slow to follow. Every man is a farmer and is directly dependent on the land. Even the blacksmith is only a blacksmith as a sideline.

Yes – I agree the fly problem is a large one. I am dashed sure if I were a DC I would improve it in my district immediately, by the ordinary method of sending sanitary labourers, trained for the job, round the district. (69) I would make the Chiefs dig latrines and see that they were used. If necessary imposing heavy penalties where, in trekking, I found too many flies in a village or garbage round compounds. Ordinary MO's are supposed to combine the office of Sanitary Officer, where there are none of the latter. But either the MO has so large a district that he seldom has a chance to leave the hospital or HQ; or just time to travel along motor roads to dispensaries in such places as Zuarungu (now that the hospital has been moved to Navrongo) and Bawku. He never sees conditions in the bush, only the patients who came to him. Either this or else he is too angry in his attitude towards Government, for sending him with all his qualifications to a badly equipped place in the Northern Province. (70) He resents being sent to a lonely station, no ice, electric light or lucrative private practice, which his colleagues in the wealthier districts in Ashanti and the Colony enjoy. The trouble is that it is difficult to get doctors to come to the West Coast, so the Colonial Office cannot select its men, and does not always get the qualified man who is a keen Empire builder.

Wilkinson, the new MO who came up to replace Sanders, is now stationed at Navrongo and is a queer bird. Exceedingly well read and well travelled with a wonderful command of the English language, invariably used on the few occasions I have met him in cynical comments about the Government. Maybe the Government's hands are tied to their sides with red tape, but all the grousing in the world will not help matters, and can only depress or demoralise one's fellows. But to return to Dr Wilkinson, he has a most unsanitary red beard. It wouldn't matter so much if it wasn't red, and was a really decent growth, but it is very straggly and thin in places. It makes him look like one of the apostles. Symond on the phone was more to the point and said like Jesus Christ. We always seem to have queer doctors up here. If they are not they don't remain. I was sorry Quigley left, two days after I arrived. He was an awfully good bloke I believe. But enough!

Locusts have occurred in small sporadic swarms this year, but are well within the limits of local control, and I do not think will be serious this year. It would affect my job up here if they were, since the staff would have to go out on patrol.

P.S. You must buy the April number of that foul magazine 'Britannia and Eve' to learn just how we live in Tamale. It is written by a nasty piece of work called Arthur Mills. We gave him a game of polo in the tournament at Christmas time, and he fell off his pony. The length of time since he last played changed three times in the same evening. His friend is the Commissioner, that arch blighter Duncan-Johnstone. 'Where I spent my Christmas' is the guest house behind the DC's compound. 'George' is Major Lionel Hagit – an awful ass. The reason why the subalterns left the Mess when he came in was that they simply loathe the sight of him. It is interesting to know that we all keep black women. (71) I wonder if this sort of stuff does any good. It can't do much harm and I suppose it is what the public want. We are all furious and amused and only wish we had taken the opportunity of kicking Mr…. Arthur Mills when we extended hospitality to him."

Roger Page, the District Commissioner, went on leave towards the end of June and I was left as monarch of all I surveyed. It took a little getting used to being entirely on my own. Reading was important and generally on Saturday evenings I would read a play by Shakespeare, Bernard Shaw or Somerset Maugham. I was fond of playing a ukulele to myself in the evening and going over familiar tunes, and the gramophone too was great company.

12th June 1932 "Mother dear,

My garden boy has accomplished more since half past five this morning than he has done for a fortnight. I have cut my hair, made up my accounts, and it's only 10 am. The peacefulness of everything! The cat sleeps peacefully and everything seems for the best in the best of all possible worlds.

Page is away spending the weekend at Navrongo. I think I am going to enjoy being on my own, and I'm sure I will look back on this spell in Zuarungu with pleasure. As Symond said in his letter last week 'If nothing else comes of it, you will never forget the experience, and you will at least have the feeling that you know some places thoroughly. How many of us can say that, after years in the country with our noses held down to some factious scheme, aimed at a petty report?'

Since my last letter I have done a week's trekking in the South Eastern district, where the population is spaced out, and there is some very fertile land. I adopted Page's scheme for trekking, starting at the crack of dawn 4.30–5 am and riding through the morning about a quarter of a mile ahead of the carriers.

Everything is very green now, a great difference from when I first came up. It was a delightful trek, with hills all around, and the Gambaga scarp always south of me on the other side of the Volta valley. I identified a lot of trees and made some progress with locating villages. I am endeavouring to produce a map of the district showing the distribution of compounds. This will be useful in discriminating the different problems in the various districts, which I think the yield figures and observations will bring out later"

28th June 1932 Rest house Navrongo "Dear Mother,

You will note that I am in the relatively highly civilised spot known as Navrongo, which boasts four white men in addition to the Fathers at the Mission.

I set off early this morning from Zuarungu, walking, and had arranged to change carriers and breakfast at Sambruno, which is about half way – the whole distance being 24 miles. However, as luck would have it at Bolgatanga, four miles from Zuarungu, I found a lorry which had got stuck on this side of the Pwalagu drift when it was washed away. It was going to Pwalagu to pick up some loads for Navrongo. So while it was going down to Pwalagu I had breakfast in the Rest House at Bolga, and came on, handsome like, for 1s a mile. I lunched with Orr the Vet, and an excellent game of tennis with Orr, Oliver and Father Robert from the Mission, and tonight dinner with the doctor. I take back the unkind things I may have said about the doctor, for while he grumbles and dislikes the country, he has not the slightest intention of remaining. Therefore, he does not belong to that class of Government Servant, who grumbles but has every intention of remaining, and draws as a matter of course a big salary for faithful inefficiency.

The doctor is an astronomer. I have been looking at Saturn just now through his magnificent three and a half inch telescope, and also at closed and open groups in Scorpio. My head whirls with alphas, betas and many technicalities. I have also been introduced to the nautical almanac and astronomy charts, and find the subject both fascinating and interesting. In addition, the doctor has very advanced views on diverse subjects. It is indeed pleasant to find somebody sufficiently scientifically biased in their outlook with whom one can discuss diverse subjects and philosophies. I must admit that many parts of our conversations were lost on me. But it was a most interesting and elevating evening. The mushroom omelette and anchovies were excellent, in spite of the doctor's apologies for the meagreness and quality of the repast, neither of which criticism could I agree with.

My main objects for honouring Navrongo with a visit, apart from a little outing for myself, is to cash the monthly vouchers at the Treasury. I can check the balances, this being the end of another quarter, and telephone John Symond tomorrow morning. I sent him a wire today warning him for duty, between the hours of 9 and 10 am tomorrow morning.

I am leaving on Thursday, and the doctor has very kindly insisted on running me back as far as Sambruno, where I propose to spend the night and see how the work is progressing there. I did not need persuasion to accept this kind offer, as it meant 12 miles walking along a motor road, which is not much to my taste. Give me the bush path.

The work continues happily. Early millet is being harvested, and in most districts local opinion is that they are the best crops for the last few years, which is all to the good. Also the plot arrangements and system of recording, which of course had to be formulated before the crops were growing, seems to work according to schedule. It will, I think, yield valuable data which I require, which is a pleasing feature of things at the moment…"

There were many mosquitoes in Zuarungu station during the rainy season, and I became aware of the source of most of them when I was surveying farms close to the station.

14th July 1932 Zuarungu "Dear Mother,

It is an indescribably lovely night and I would that you could share it with me. An amazingly clear sky, a bright moon, one can almost read by it. The Tong Hills six miles away wreathed in mist, and a rare cool breeze blowing. Everything seems still, although if one listens one can hear in the distance some messages being shouted from one compound to another. It started earlier in the Chief's compound, and Abugari tells me that they are saying that nobody must have water standing in pots or calabashes in their compounds, or they will be taken to the white man.

This latter is the direct outcome of a practical demonstration to the Chief this morning. I hatched out a tremendous number of mosquitoes from a little water I collected earlier in the week, when I was visiting one of our recorded compounds to weigh some sample plot yields. It was an eye opener to me, and I think the Chief fully appreciated the lesson. I must confess that since I started my campaign a little while back, against mosquito larvae, there has been an instantaneous drop in the number of these pests around my house. They were chiefly Stegomia, which breed in old tins and bottles and never fly far, and carry yellow fever. The anopheles mosquito carries the malaria vector.

I am negotiating with the doctor at Navrongo for a couple of sanitary labourers to do a little draining round the station, to clear ponds and waterholes. This latter job would be a normal practice, of course, every year with ponds at home on the farm in the non rainy season. It gives me pleasure to put into practice a few of the oddments I have gathered here and there. Being on my own I can do it without fear of insinuating that the DC or MO is not doing his job properly; and of course, as a result my health and that of my boys should be better and efficiency maintained.

I have just finished dinner and am writing this to pass away half an hour before retiring, which I usually do between 9 and 9.30.

Gibbs, the DC at Gambaga (72) has been over recently, he is also DC of this area of course. I have got a certain amount of pleasure catching him out on one or two points. Every DC I have met so far says 'of course these people are terribly hard working' and alas that 'they are short of food because they are short of land'. In the first case I took Gibbs out on a grand farming morning, after asking the density of population I bet him (he would not take me on) that on any farm we liked to go to at 10 am, we would find plenty of useful hoeing to be done and hardly anyone out doing it. I said that we would go into the compounds and find the men lounging around chewing fresh millet. Such was the case. (73) As for the land question I simply have to say 'are you quite sure, because that is what I am up here to find out. If you know it to be a fact, then I can save a lot of time'. They invariably rise to this! The trouble is that these and similar generalisations are put into reports etc., and you have got, (or rather I have got) to start off by disproving many accepted facts. (74)

However, this is a case where figures will speak, I think. Taken too often as they stand, and not checked against the facts. Then they are multiplied (and any error) by the population of the district, and produce some perfectly amazing results, which in their turn will be quoted higher up the field anyway.

My job has, as a matter of fact, enlarged considerably in scope. It should be of the greater value in Assessment for Direct Taxation. Gibbs very decently lent me his file on the subject of Direct Taxation, and some of Major Jackson's (Chief Commissioner) memorandums and letters are ludicrous on agricultural matters. If it was not for the fact that the tax will be very low initially, considerable hardship might be imposed, if the only information for assessment is what I saw in the file. However, a confidential circular has just come round, urging co-operation between Political and Agricultural Offices.

This gives me great pleasure, as this idea of water-tight compartments and unnecessary duplication of effort, has given me thought for some time back. Also as far as Direct Taxation goes I am all for it. The Chiefs with their own Treasuries will then have some power, and will be able to do things. It will be under the guidance of the white man, anyway at first, and if applied properly it may provide the necessary stimulus for better farming in the area. I am very afraid though, that the tax suggested at the moment is going to penalise the industrious farmer.

The transition stage, as always, is awkward. One cannot obtain free labour from the Chief, and Government has not the money to pay the wages of hired labour. The Native Treasuries will not be up and working for some time. They will pay labour for social services like road maintenance, repair and provision of drains, ditches, latrines etc. As seems usual in this Colony things are being done in any order, very ill considered and will probably be rushed at the last minute. This is so that the Governor can put it in his speech, when he opens Legislative Council, as being done.

But I am rambling. Don't take the above too much to heart, it will probably be contradicted next mail. I am only just having my little grumble.

Here we are lacking rain. Early millet is in and yielded well, but we have only had just over one inch so far this month. The mean is over 6" and the guinea corn; late millet and rice need it badly. I am given to understand, on good authority, that this lack is due to a deadlock locally between the 'rain makers' and the 'Chiefs'. (75) The former are holding out for gifts of goats from the Chiefs, and the latter are dammed if they will give them any goats. So far the rainmakers are laughing. But I have little doubt an agreement will be reached soon. I hope so because I want to go on with planting groundnuts, cotton and Bengal beans on my plots. From another point of view the weather is magnificent. Bright crystal mornings, and the fiercest rays tempered by the prevailing breeze from the S.E. I never get tired of the sun, and always miss him on a dull day. Temperatures keep down in the 80s, and although I have no hygrometer here I should think the humidity is not above 60%. The water table is still very low…"

I have been appointed by the Medical Officer as Mosquito Inspector under the Ordinance.

2nd August 1932 Zuarungu "Dear Mother,

It is a confounded nuisance having the post office 24 miles away and only a messenger on foot. He is leaving at dawn tomorrow so I should get this to him in ample time.

I got back from trek at tea time on Saturday, but being the end of the moon, my monthly report took me longer than expected.

They have been having trouble over at Navrongo where Oliver is DC. In a short space of time two little girls have been found beheaded and their bodies ripped up. Some juju palaver, in which the Chief (the Paramount Chief of the Builsa tribe) is mixed up. These are the people to whom we are giving indirect rule. No one of course will say a word, and Oliver's court interpreter is also implicated.

Placid is still the most concise idea of my existence and if you want something a little fuller, one might say – pleasantly placid. I sometimes wish though that I had done something which weighed on my conscious, and made me want to banish myself entirely from my fellows. It is certainly an experience I would not have missed for anything.

I find it difficult to refrain from applying what I know to be good farming practice elsewhere in criticising these people. But I honestly believe that no farmer in the world works less than he does here. They actually prefer to go hungry for 3–4 months in the year rather than do any extra work on their farms, so what can you do with such people? They have not

68

done a stroke of work since the early millet crop was harvested, and just spend all day sleeping and eating. When the crop is all eaten, out they will go and perhaps do a little hoeing on their farms to bring on the guinea corn or late millet. Then they will just eat away at that through the dry season and do no more work till the first rains. They are quite happy like that, and I don't really see what right we have got to make them discontented and work more; nothing can be done for them unless they are prepared to work more and this is the last thing they want to do. As Symond said in his last letter 'it becomes increasingly apparent that the time honoured statement of land shortage is a myth, and if it is so that the people possess the remedy in their own hands'. He considers the attitude of the powers that be will be one of 'God helps those who help themselves'…"

A fundamental matter, which early began to engage my attention, was water supplies. The wise Chief of Bongo, (76) who was rather sceptical about proposed efforts to improve farming, said that if I really wanted to do something useful for his people I should give them water. This was, though, really carried out by the Water Supply Section of the Geological Department, a small department located mainly in the south of the Gold Coast. At that time it was not represented in the Northern Territories. I realised that I was unlikely to make any progress in agricultural development until some improvement was made with water supplies. Here was a real "felt need" of the people and it was a good extension principle to give attention to matters in the order of their local importance.

Fortunately, the water table in this area was generally at between 20 and 30 feet in the valleys, and in the rainy season a good deal less. We found it was comparatively easy to dig shallow wells in the rotten granite. A shaft 6ft wide was dug about 6 ft down, and then reduced to a diameter of 4 ft to below the water table. On the ledge so formed, a wall of stones was built, either with quartz stones or laterite blocks, held together with cement. For the expenditure of a matter of £5 it was possible to make and stein a well. The Native Administration provided the cement and a mason, and the people themselves the stones and labour. So from small beginnings over a period of years the people were persuaded to construct hundreds of such wells to good effect.

It was also feasible to make earth dams in the valleys, using men to pull scrapers and women to carry soil in baskets on their heads. Very often hundreds of women would be prepared to turn out to make such a dam, and many were made. In the seepage area behind the dams useful dry season gardens were established, supplying much-needed vitamins. Fish could also be farmed in the dams. The secret of success was that the people did most of the work themselves. (77)

Rice was a crop that was not grown traditionally, but its popularity was increasing as a foodstuff particularly amongst lorry drivers and people travelling. It grew well in the valleys, and became potentially more important as water supplies improved.

4th September 1932 Tamale "My dear Mother,

I was phoning Symond from Navrongo last Monday, and he suggested that I should come down here for a few days. Amongst other things we needed to talk over my work, so that he can arrange a defence to keep me up here – for the season at least. Auchinleck apparently considers that I am wasting my time there – probably it is the first fundamental work the Department has ever done up here. It is so absurd; anyway I have the backing of the whole Political Department.

Only the other day, Dixon of Salaga, a 100 miles south of Tamale, had heard about it and told Symond that it was the first time that the name of the Department had ever appeared as useful in the Northern Territory Administration. (78) So I am not worrying.

The Gods were with me. I sent ahead to Bolgatanga and got a lorry to take me to Pwalagu last Tuesday. The river had washed the drift away and the lorry got caught on the wrong side. Another one is stuck between Pwalagu and Nasia, so wired Gibbs for it. I arrived at Nasia River by midday on Wednesday, where there is a telephone linesman. I tapped the line and got on to Symond and he said he would come out at once and fetch me. (The previous arrangement was that he should meet me on Friday, since I thought I would have to walk the 60 miles.) I crossed the river by canoe again, and had to walk only four miles to where lorries and cars can still get up from Tamale. My cook prepared tea just as Symond arrived over the crest of the hill at 4.30. We had tea and stuffed a suitcase in the back of the car, sending back my cook box, chairs and tables to wait with my other loads at Nasia Rest House, and came down with one boy. So here I am.

Captain Stewart, the Principal Veterinary Officer, now wants me to spend a couple of days with him on my way north. He has marvellous labs at Pong Tamale 18 miles up the road, and will take no refusal. However, it will be exceedingly interesting, as I am becoming more convinced of the futility of considering agriculture as apart from animal health. Jock has a magnificent bungalow, and agreed to take me north in his car as far as the road allows.

I assume that I will be left in peace, to finish my job this season in the North. I am wondering what will become of me after Christmas – whether I will be allowed to remain in Tamale. The fetish in HQ at the moment is money crops, export crops, and because of the transport problem from the NT it means development of Ashanti and the Colony. It is a policy of laissez faire up here. Food crops are not to be considered. It is wrong, of course, for it is quite legitimate to say that without the NT the Colony couldn't possibly export so much cacao.

All the labour comes from up here. (79) I doubt if the situation will improve so long as we are entirely dependent upon indirect taxation for the revenue of the country. The high prices of cacao a few years ago blinded everybody. It is now that this boom is over that inefficiency and instability is patently oblivious. But unfortunately the cost of maintaining the capital involvements, in which the Colony indulged when a surplus of 4 million existed, is now a problem when we are a half million down. But this is heavy…"

I enjoyed my stay at Pong Tamale, which was used as a station for producing the hyper- virulent cattle. They provided the serum used in the anti-rinderpest campaign throughout the whole of the country. Jock Stewart himself was a tremendous character. He weighed 16 stone and had a special bicycle with an extra bar to take his weight. West Africa provided scope for individuals with character. He qualified in Veterinary Science after distinguished service during the First World War and was capped at rugger for Scotland in 1922. He was a superb host. His bungalow was extremely comfortable with running water and electricity from their own generating plant. In the evenings it was a joy to listen to classical music on an E.M.G. gramophone with a very large horn. His steward boy, Ayidibiri, had been trained to operate it. I met the new laboratory attendant, Haynes, who had recently arrived from Britain. He explained the process of manufacturing the anti-rinderpest serum.

The Veterinary Department had a very considerable stock farm at Pong Tamale, which really made it look more like a super agricultural station than anything at that time controlled by the Department of Agriculture. It was true that the biggest problem limiting the greater use of livestock was disease. To that extent, a veterinary approach was probably the right one. However, a time was soon coming when major diseases of livestock were under control and the Department of Agriculture would be required to play its rightful part in animal husbandry.

Jock Stewart drove me up the north road as far as Disiga and I set out on foot for Walwale. Here again, I was fortunate because Charlie Orr, the Veterinary Officer stationed in Navrongo, had been up to Gambaga to look into an outbreak of rinderpest. He was on his way back with the one lorry that was still operating between the rivers. Our cooks contrived to produce a very good lunch consisting of salad and cold pigeon. We then piled into the lorry 27 loads, two cooks, three dogs, two small boys, one cattle patrol, one driver's mate and two Government officers to proceed to Pwalagu ferry on the Volta River.

When we got close to the river we found that a great flood had come down since I had crossed a week before. It was necessary for us to wade a quarter of a mile through a lagoon, to get to the riverbank where the ferry canoes could reach us. What started off as a gentle walk through shallow water turned into a walk through water up to our necks. Our treasures were placed in haversacks and carried above our heads. We were not in a position to worry too much about the possibility of crocodiles in the water, but there were leaches. In due course we got to the riverbank and were able to raise the ferry boys on the other side. Charlie Orr's "A" model Ford, which had been left at the rest house half a mile from the river, started at first pull.

It took us back to the river where our boys had very sensibly sent over some loads containing dry clothes. Once across we made our way to the rest house, only wishing that all the staff at the Secretariat of Accra could be made to trek north in the rains. They would then realise how official correspondence and wires to "Please expedite returns so and so" rankle in the bush. It seemed incredible that this same river, when I used it in February, was a dry sandy gully with a firm embankment road across it. It is a pity Auchinleck has got it into his head that I am wasting time up here, or rather that it is a waste of time my being here.

11th October 1932 Zuarungu "Mother dear,
As I have said before it is good to be back in the pleasant tranquillity of Zuarungu – ah I hear a 'kura' – hyena (makes a noise not unlike the first

part of the E-Aw of a donkey). Crops are going well, but alas I find all the paths on my observation plot beautifully weeded and soil mulched, and many plots badly wanting to be weeded. This is after carefully explaining to my overseer, Yakubu Singa, that I wanted the grass and weeds to grow on the paths to check soil erosion. It is rather typical of your efficient African Official – a man getting £11.10s per month who has risen in the Department without a check since 1914. They are hopeless. It is European Supervision in this country which makes the most footing jobs so expensive. I am absolutely convinced that if we left this colony today it would show no signs of our intrusion in less than 20 years. And perhaps the native would be happier. I often wonder. (80)

In the meantime, work progresses satisfactorily, the guinea corn and late millet are flowering and setting seed and should be ripe in another month. Groundnuts are being lifted and within two months all the figures should be in from my 300 sample plots; it will then remain to sort out, analyse and come to some conclusions to be embodied in a report.

Symond sails tomorrow from Takoradi and old Cowan is back, but would you believe it, he has been sent to the Rice Factory at Essiama down on the Coast. Symond is of the opinion that it is just to spite him. Steel is taking over from Symond, but will not be in Tamale before Symond leaves. Steel has never been in the NT in his life before, and is simply being sent up here to get him out of the way in Head Office.

Things do not look too bright for poor old Tamale Experiment station. For the last three seasons a different man has harvested the crops to the man who planted them and watched them grow, and how the devil an experimental station can hope to run and produce results of any value heaven knows. If it is not going to be run properly why not close it down? It's all cacao, cacao, cacao simply one or two wealthy natives in Ashanti, and the Colony have made their money from it; they have sons qualified as lawyers, who are given the run of the native press. One is almost forced into that bad attitude of mind of 'why worry', our salary gets paid just the same…"

14th October 1932 "Dear Mother,

'Rana ta kana wata, sariki yana so ya bugga da wasa' 'The sun is catching the moon, the Chief wants to beat drums and play'. Abugari gave me this interesting piece of information as I finished dinner, and of course I graciously gave my consent. So the night is hideous with caterwauls and drumming and as I sit the moon is practically eclipsed. (81) Since it is as visible at Greenwich, providing it is clear, you also may be looking at the same moon at precisely the same time. I doubt if the respectable residents in Beckenham will get such a kick out of it as the simple Fra-fra; this will be a grand excuse of opening a bottle of '75. I've no doubt that the pito will flow freely in the Chief's compound tonight, and tomorrow my labourers will be dopy.

A veritable cacophony of noises is being given forth. Michael barking and Anasuri, the small boy, making the devil of a din with a kerosene tin behind the compound – it reminds me of 'All Clear' in the air raids. The noise has quite suddenly stopped and the night is strangely quiet again – the eclipse has reached its peak and lightening us again. I have no doubt that there are many weird apparitions associated with the phenomena, which I must try and find out. Abugari seemed to quite realise that it was something to do with the sun…."

During October, I paid a visit with Charlie Orr, the Veterinary Officer, to the French station of Po across the border north of Navrongo. We took Brother Aidan from the Mission as interpreter. Orr wanted to give some information regarding the proposed visit of a French veterinarian to Pong Tamale. At Po we met a vivacious Frenchman who was in charge of the Armistice Celebrations to be held at Ouagadougou, the capital of the Haute Volta Colony 120 miles north. Apparently, the French were going to hold a race meeting on November 11th, and a number of what they called "fetes" and wished us to attend.

So Jack Olivier, the District Commissioner Navrongo, and I were to visit Quagadougu, going up in his open Ford, which had no windscreen or hood. Unfortunately, an officer of the Regiment and a missionary had died during the previous week from yellow fever, and there were to be no fetes. However, a race meeting was held in which Jack Olivier took part wearing mosquito boots. He rode in the Grand Prix de L'Armistice and very creditably came second. We were astonished to find so far inland such comfortable bungalows, with electricity and running water. We noted that the bedrooms were fitted with punkahs, over the beds. It was all very different from our own set up in the Gold Coast. We returned feeling that the trip had been very worthwhile.

It took about a month to complete the harvesting of some 300 sample plots throughout the district and to gather the yields from the various observation trials. I left my African staff to carry on in Zuarungu during the dry season, and took my data to Tamale to write up the annual report. My Director had suggested that this could be published in bulletin form, but I was uncertain of the wisdom of doing this as a result of only one season's work.

7th November 1932 Zuarungu "Dear Folks,

Yes, the star atlas arrived but I regret to say I have scarcely opened it yet, I have been very busy and am in the thick of things now. My harvesting is going on. I need to be in exactly five different places at one time, at once and 12 miles apart. It simply cannot be done, so I am staying in one place and employing a multiplicity of messengers to bring in reports of progress from each centre every day. In this way, I get some measure of control. I laid my plans as well as I could, and trained the staff so I can only hope for the best. I am powerless now to correct any mistakes. Harvesting of my scattered sample plots feels like a heavy steam roller too heavy to be steered.

My latest brainwave is to establish model unit Native Administration Farms under each big Chief in the district. They could be managed entirely upon the lines I consider farms here should be managed. These need not be an expense to the Government. They could form a most valuable channel for extending the results from experiment stations directly into practice in the countryside. The people will work the farms and reap the profit. We will have invaluable records and our fingers on the pulse of the countryside. They could also form useful centres for the bulking and distribution of improved strains of seeds. In this way, our results of experiments would have more than academic interest. But Auchinleck has other views; he says that the function of a Department of Agriculture is to give government sound advice on agricultural policy. I maintain that its function should, in addition, be to execute and carry out that approved policy . Its success should be measured by the number of farmers forced or persuaded to adopt improved methods…"

It was exciting to be back in Tamale, and to enjoy running water and a long bath after only a hip-bath. I spent part of my time helping with the harvesting of the experiments on Tamale station, and then after breakfast worked in my own office on my Zuarungu data. This took me happily up to Christmas time and for the third year running I enjoyed the festival in Tamale. I was lucky to be there for the annual polo match against Accra on 2nd and 4th January, which we won.

In February, I took local leave and drove down to Kumasi, and then by train to Sekondi, where I was met by Gordon Cowan. That evening we went sailing round Takoradi Harbour and circled the M.V. Apapa homeward bound. We fished in the lagoon behind Essiama. Gordon, who was a keen and expert fisherman, and had recently hooked two barracuda weighing 31 lbs and 20 lbs. There was excellent surf bathing on a lovely sandy beach at Essiama, and intriguing mangrove swamps on the lagoon. On the whole, it was cooler at Essiama than it had been in the Northern Territories. There was no doubt that I benefited after 14 months of my tour, always a critical time.

Whilst at Essiama a shattering telegram arrived from "Agrestic", the telegraphic address of our Headquarters in Accra. I was told to report to Kumasi immediately for special duties in North Ashanti. (82) The blow had come; I was to be removed from my favourite Northern Territories. This meant that our trio was completely broken up, Symond was going to Koforidua on return from leave, Cowan was here in Essiama and I was to go to Ashanti, so much for continuity. However, this was to be a new experience for me in a new province. I reported to the Provincial Agricultural Officer in Kumasi and learned that I was required to develop a groundnut scheme in North Ashanti.

Apparently, the United Africa Company had arrived at a gentleman's agreement with the Chief Commissioner Ashanti, that it would purchase all the groundnuts offered for sale during the coming season at a minimum price in a defined area. Some 15 tons of groundnut seed was imported from Nigeria, with the seed coming up to Ejura, the headquarters of the scheme, on the new road. The snag with the scheme was the comparatively low price that was offered, £14 per ton in shell.

At Ejura, (83) I lived in a tin- roofed shack, without doors or windows, located on an old agricultural station planted with mature but neglected mangoes, limes, lemons, oranges, grapefruits and tangerines. At the weekends for a change I went three miles down the road to stay in the magnificent rest house at the top of the escarpment, which marked the southern end of the Voltaian sandstone region. From here one had a wonderful view over the forest country to Mampong in the south 30 miles away. The rainfall in Ejura was 60 inches a year falling in a distinct wet and dry season giving savannah vegetation. The soil was red sandstone and the people lived in compact towns and villages scattered at intervals along the main roads. My district extended for 90 miles north to Atebubu and from there 40 miles east to Basa in Brong country. The road went on to Kete Krachi, where my old friend from Zuarungu, Roger Page, was now District Commissioner. On one occasion, I went through to stay with him, and we went for a ride in the old Rolls Royce that he brought back from leave.

18th March 1933 Ejura on Transfer "Dear Mother,

I have been exceedingly busy and covered many hundreds of miles this month. March is the end of the financial year, and one is up to one's eyes in vouchers and overspent notes.

I came up here with Kinlock on the 7th, having spent a pleasant enough week in Kumasi sorting out my kit. Half of it I had to be put in store, as I am back again in bush quarters. We are operating along about 90 miles of roads from Ejura, which you will see on your map some 61 miles to the North of Kumasi, up to Atebubu and from there west to Bassa on the Krachi road.

Kinlock, who did the spade work last month, is a pleasant Scotsman with stacks of energy. He is an Inspector of Plants and Produce running Co-operative Societies in Ashanti. I came up in his car and have had meetings in all the villages along the road. The idea is to guarantee a minimum price, and to give an unlimited market for groundnuts. (It has been suggested that Whipsnade is responsible for the demand.) Unfortunately, the price which can be offered is very low. Personally, I don't see much hope of an industry becoming established for a little while at least, until the people have got used to a small profit on a big turnover. However, it is a new part of the country.

We are just out of the forest belt on magnificent deep alluvial soil, receiving a rainfall of about 60 inches p.a. Many of the villages are Northern Territory native settlements – Zongos, (84) as they are called – in which the language spoken is Hausa, and it is from the NT population from whom we are likely to get the best response. I am very pleased to find my knowledge of the language is of very definite practical value. The other villages talk Twi – the Ashanti language.

My house here is of native construction but has a corrugated iron roof, and is really not at all bad – no furniture, but with my bush stuff of course; however, I get a fiver a month allowance in lieu of quarters. Ants are the chief trouble, they eat into everything. It is the old Garden House. There used to be a station here way back, and there are some fine limes, mangoes, lemons, oranges, grapefruits and tangerines in the garden which will all be ripe very soon now. The DC is at Mampong 26 miles down the road, but there is a road foreman somewhere along this stretch; however, I don't suppose I will be troubled with many callers…"

By the end of April, all the groundnut seed was distributed to some 1600 farms in 50 villages. The seed was issued on loan, with the idea that if the scheme was successful the seed recovered would be used to extend the industry to the Wenchi district. Most of the villages were in two sections. The smaller native section was the local tribe, where the people spoke Twi and were reluctant to come into the scheme. The larger zongo section comprised of strangers from elsewhere, and had Hausa as the lingua franca. It was interesting that the response was much greater on paper from the zongos, but when it came to the acreage planted the indigenous people proved much more reliable and effective.

I spent a weekend in Kumasi when my old first car required attention from the garage of Allen & Grouchy. Then in May I paid a visit to Mampong, 26 miles south of Ejura down the escarpment and into the forest country. The occasion was a visit of His Excellency the Governor, Sir Shenton Thomas, accompanied by Lady Thomas and their daughter, with Captain Lotinga of the Regiment ADC. We had a pleasant dinner held in the open. I was able to give up- to-date information on the progress of the groundnut scheme, and also of the survey work in Zuarungu. The discussion showed how fundamental the work of the Department of Agriculture was to the development of the country and how basic to administration.

At the end of June 1933, I went on leave on the R.M.S. Apapa. I bought a new Ford 8 while on leave, which I took back to West Africa.

Agricultural Officers House, Zuarungu, 1936

DC Lt. Col. G.H. Gibbs and DC
D.N. Walker in Navrongo 1934

Dr Meyer and Sonia Fortes at
Tongo 1934

Lynn's House staff, Anasuri and
Asubiri Fra Fra 1936

Anane Fra Fra, Bongo Na. 1935

Chief of Zebilla 1938

Compound of the Chief of Zuarungu 1934

Pwalagu dry season drift 1934

Chapter 4 - 3rd Tour 1933–35

Mixed farming, the Arrival of Dr. Fortes.

I sailed back again in December 1933 on the Apapa, with my Director, G.G. Auchinleck, and Jock Stewart, Director of Veterinary Services, as fellow passengers. We had many talks about the role of the Agricultural Department in the Northern Territories and the need for mixed farming as being developed in Nigeria. I was relieved to be told to proceed to Tamale for instructions.

On my way through Kumasi I learned that the groundnut harvest in North Ashanti had been a good one. Alas, the price of groundnuts on the London market had dropped from £14 to £ll.10s a ton between the time of planting and harvesting. The United Africa Company could not afford a price which was satisfactory to the farmers; in fact the growers found a much better market in Kumasi and adjusted their acreages in subsequent years.

Times were difficult in the height of a world depression, but this situation was hopeless from an agricultural extension point of view. I placed on record that I should never again be sent to Ejura district to encourage the farmers to do anything. This question of the price, to be paid to growers for primary commodities on world markets, is one that has bedevilled development. In 1933, the cacao crop for the Gold Coast was estimated at 218,000 tons, but there was still 78,0OO tons in store. The price for a 601b load of fermented cacao was 3s whilst in the boom of 1923–26 it had been £3.

Upon my return to Tamale I was told of the change in Government policy regarding our work in the Northern Territories. The old team of Symond, Cowan and Lynn was to be reinstated with an additional European officer to be sent up when available. This meant that the work could continue in Dagomba centred on Tamale station. I could return to my beloved Zuarungu, and establish an experiment station, and do survey work in Navrongo and Bawku districts.

"The Beetle", Trevor Lloyd Williams the Economic Botanist, (85) continued his cotton selection work and extended his interests to guinea corn. He was migratory, being based on Kumasi and coming north for the harvest season. His help and wise observations were much appreciated. I spent a relatively quiet Christmas, my fourth in Tamale; there were dinner parties at the Regimental Mess with "the Beetles", the Blairs (DC Dagomba) and Eustace Rake and Muriel his wife.

7th January 1934 Zuarungu "Dear Folk,

I have to ascertain roughly the distribution of Acacia Senegal, from which a particular kind of gum Arabic is obtained, and I have a record from 1933 to say it occurs round the Tong Hills in which direction I propose to trek.

It is grand to be back in my old house again, although it has been allowed to fall into a sad state of disrepair. It is, however, to be renovated, including re-roofing which will be rather nice, but an awful mess until it is all finished.

I saw the New Year in up at Bawku with Ellison the DC there. I arrived up there last Sunday and was apparently a bad juju for just as we were sitting down to lunch, a messenger dashed up to say that there was a riot in the Bawku Naba's tribunal. We rushed down and it took Ellison nearly half an hour's shouting to settle it. On our way back to the bungalow, we saw flames leaping up over the town – a terrible thing in this dry harmattan weather, with a gentle North breeze and densely aggregated houses with thatched roofs. The police were turned out and the whole place in uproar. Everybody shouting but doing nothing about it!

Fortunately, there was a lull in the breeze, which gave us time to pull the roofs off the houses nearest the burning compounds. We then had the mud walls pushed in, onto the mouldering wreckage. Nobody was hurt, which is truly remarkable considering the tiny little hole low down on the ground which serves as the only door to each hut. Only about 50 houses were destroyed instead of 5000. It was bad luck on the poor devils whose entire possessions went, but it won't take long to build again. Hoarse with shouting and smoke, and very hot and grimy Ellison and I eventually sat down to lunch at 3.30 pm instead of 12.30…"

"…11th January 1934 Shega Rest House "Dear Mother,

I have been 'on trek' now since 6 am on Tuesday, when I left Zuarungu. The men of Bari (Ba'ari) and Zokko were going to re-thatch my house, and since I wanted to get in touch with the district again, I decided that the best thing to do was to take some walking exercise. The aches in my shins and thighs tell me how good and desirable this is. Needless to say I am feeling very fit and well. Trekking always has a tonic effect. I think, even if nothing else happens, a 12 mile walk in this country is a day's work. Then there is a spirit of adventure about it, and great independence, which appeals.

There are little hills all around here which makes the country most interesting. Then I am interested in trees, and I am quite gratified by today's trek to find an abundance of Acacia Senegal. A firm in Accra wants a quantity of gum Arabic from this particular species. Whilst the price they offer is, in my opinion, a low one, anything that brings money up here is a good thing. Actually, I missed the road this morning. I should have been in a place called Kinigu, about seven miles south of here over the Volta River.

As it turns out it is a good thing, for I would otherwise have missed a messenger from Symond sent up from Tamale . Stalking on ahead in the cool of the morning, I missed a fork road, and didn't stop until I saw three oribi (little antelope the size of a goat and grand chop); I stopped and waited for the first carriers to come up to me (they had the cook box for the breakfast palaver), and I sent one back for my rifle. He brought up the next carrier, who had my shotgun which was no good. Then my cook appeared and only three loads, there are 12 in all, and no Abugari and no Mr Yendi (my overseer and interpreter). We breakfasted and still no sign of the rest – they had taken the right road. As I was at breakfast I was amazed to see coming along the path, through half- burnt grass and trees (for the place was miles away from anywhere and anybody) – Grunnah, Symond's garden boy, bearing a letter from Symond telling me that Professor Shepherd, Carnegie Professor of Economics, from Trinidad, proposed visiting me on Saturday for a couple of days.

This was very unexpected. I knew Shepherd was in the country to advise on the Co-operative movement. Arriving in Shega, a messenger was sent to intercept the rest of the ménage, and they trooped in at noon. Tomorrow I trek back to Zuarungu via Tongo, and hope the roof will be on my house, or else Shepherd will be able to live in the round rest house and tell his friends about 'roughing it in the outpost'.

Shega is Abugari's home town so I am persona grata here. The compound seethes with the biblical offerings of the countryside – fowls, eggs and milk and just at the moment a leg and liver of a sheep has appeared from the Chief's compound. Shega seems to supply half the servants of the country. Sando, Cowan's boy, comes from here and the familiar face of Aborilea, one time very small boy and now with an officer in the regiment on leave. Then the guardian 'fowl boy' in Tamale appeared all grins and bearing a chicken and eggs. I suppose their local prestige goes up by having such intimate contact with the deity! Then they congregate in the kitchen for a yarn with their old mates. I expect they discuss the particular shades of purple that the young master's face or language was when they dropped the gravy boat, or broke the soup plate in the 'good old days'. They are very amusing and keep dodging off to their houses to change their raiments, appearing now in their late master's old bush shirt, then in a marvellous multi- coloured shirt bought on a visit to Kumasi, and so on. But it is all very amusing and gives innocent pleasure to everybody concerned.

There is still a touch of the harmattan about the nights and mornings are joyous. I sleep out under the stars, and am grateful for two blankets. Awake at the dawn, or a little before, by the noise the small boy makes getting the kitchen fire going. Then a large pot of coffee and a biscuit while the loads are being packed up, and I usually manage to get off by six or soon after, wearing a sweater until breakfast time. Even then it is quite pleasant and no sweating till about eleven. They are cheerful ones, the Fra-fras, while everything is still; as I write one can hear the hum of talking and singing coming from compounds, scattered about over the undulating country. Each one has its little blazing fire made from guinea corn stalks.

Some of them are perhaps talking about the long streak of white man up at the rest house who has a bee in his bonnet about yams and things. These ordinary trees are another particular madness of his. He is a funny man too — When we say 'We don't grow yams, because our fathers did not grow them,' he says 'because in some years, such as when locusts come, your fathers went short of chop that is no reason you should.' (86) As if we could control 'wuni' and the comings and goings of locusts (this argument never fails to draw many 'Aka's' from the crowd and much tittering). They are a pretty hopeless but cheerful people, one cannot help liking them…"

The country was pleasant for walking or riding being open farm land with scattered trees of which I had already identified around 60 species, all of some value as feed, fibre, medicine flavouring or timber. I particularly wanted to locate a concentration of Acacia Senegal as a firm in Accra was prepared to pay 26s a cwt (hundredweight) for the gum. I was a bit sceptical because the price seemed low for the labour involved, and it was unusual for a commodity of vegetable origin from wild sources.

It was my custom to start walking in the early morning when it was cool, and to breakfast under the shade of a tree at about 8 am. My cook was experienced at picking a good site, and prepared a full breakfast of porridge, eggs and bacon, or kidney on toast, sometimes chicken in a mustard batter. This was followed by Cooper's Oxford marmalade and ryvita or rolls with excellent coffee from Ashanti. After I would smoke a pipe of tobacco and read a tale from Blackwood's Magazines. It was a lovely life.

On these occasions, we travelled self-contained with 20 carriers carrying head loads of up to 40 lbs made up of camp bed, bath, chairs, table, lamps, cook box, crockery, stores, office box and so on. The reward was only a penny a mile but there was never any shortage of recruits, indeed many carriers would elect to go the whole way perhaps for three weeks, or they could be renewed post to post. (87) Rest houses, a group of round mud thatched houses, were strategically placed 10 miles or so apart and it was the Chief's responsibility to keep them in order and supplied with water and firewood against one's arrival. Generally, the Chief met me on his boundary and escorted me into the rest house, then after greetings and local news a discussion on the purpose of the visit and the problems of the area. They were charming, hospitable people and very tolerant. Whilst one did not expect a great deal to come of such a meeting it was important to establish mutual goodwill and confidence; personal contact paid dividends.

6th February 1934 Winkgo "Mother dear,

You will gather that I am fond of travelling in a dignified and leisurely fashion round the district, and you would be perfectly correct. It never fails to have a tonic effect – both physically and mentally. I do like to sit in Zuarungu with a definite picture in my mind of the district and its people, which can only be obtained in this manner. Also I am having the floor and plinth of my bungalow re-swished and beaten in my absence. So with a new roof, fresh floor and the mason working for a month on outside repairs and cement wash, my house will be fine and waterproofed for the rains. It only remains lime washing the walls inside now.

I have not had a chance to commence this letter earlier. There has been so much activity in the ordinarily so serene and placid Zuarungu, that I have been utterly lost. It is only now after five days walking, during which I have covered not less than 65–70 miles, that the nerve ganglions have stopped quivering and settled down to that steady normality which suits the job in hand. A word of explanation.

Old man Cowan duly arrived as per schedule. We had a grand week together, which went all too quickly. I declared a public holiday for the European staff of the department in Zuarungu, from all but routine jobs. We derived considerable benefit from stimulating discussion on agricultural and other matters. On the following Saturday Symond arrived in time for lunch, bringing a lorry load of yam seeds for me – at least that was the official reason. He brought three bottles of champagne, which we consumed for dinner that night. It was certainly no ordinary occasion, for it is many months since we three were gathered together in one place. Right royally we celebrated until the third hour of the following morning. Alas, Symond and Cowan departed after breakfast, which for obvious reasons we took apart. Just as the Talbot snaked away down the drive His Excellency The Governor, appeared from Bawku in the opposite direction.

I was called upon for an audience, and to my pleasure and surprise His Excellency remembered meeting me at Mampong when I was at Ejura. It was very informal, our conversation taking place over a mug of beer in fact. I was, naturally, a bit nervous at first, but soon found myself arguing with the old man and receiving an invitation to drinks in Navrongo, where he was staying the night.

The same evening, Sunday, I was consuming a modicum with Colonel Gibbs and his Mrs, who spent the night in Zuarungu, when a Dr and Mrs Fortes (88) rolled into the station asking for me. They were bearing letters of introduction from my Director and Professor Shepherd. They are queer and rare little birds of tremendous knowledge and intelligence, along their own line – Anthropology. They are out on a tour from the London School of Oriental Languages. They propose making Zuarungu, or some place near, their centre for an intensive sociological study. They are terribly keen and earnest – Oxford and Bloomsbury. He is a South African and she Russian.

I will no longer be alone, the uncrowned king of all I survey. But association with them will be exceedingly interesting and valuable to me. They are very keen on my work; indeed, I was amazed they heard of it in Accra from outside sources, so they went to Auchinleck. (89) One is very apt to imagine oneself on an island, with no connections with the outside world up here – but little does one know. It is suggested we might do something jointly. The technical agricultural aspects of their work are all important to a full understanding of these people. To what extent I will be able to do this, however, depends a good deal on staff. Already I am very short and have a very ambitious programme for this season. We are working, also, on two different systems for rather different reasons. I have taken agricultural records on the farms of families taken at random over the whole district, whilst they propose to select a limited area (perhaps two or three square miles) and work through all the families within that area. The ideal would be if I could do an agricultural survey of the same area. But we may yet divine something. Anyway, they arrived in the dark on Sunday night with lamps and beds still packed in boxes as by Griffith McAlester. What sort of a place they thought Zuarungu was and at that time of night, I cannot imagine. It was a bad and unpopular start for they were quite unannounced. (90) Anyway, Gibbs gave them dinner, and I fixed 'em up with lamps and meals the next day till they got settled. They came into Navrongo with me on Monday morning to meet the Governor. Whether their enthusiasm will outlast the rainy season remains to be seen. But in the meantime as from the blue, I have some interesting and stimulating neighbours for a year.

I'm afraid my letter has been rather full of the Fortes, (91) but necessarily so. This country is terrible for discussing personalities; perhaps because of a limited number of other topics. Other folks are forced very much upon one here, as they would not be at home – sometimes a good thing and sometimes the reverse. I can see a good deal to be said for the comparative seclusion that is to be obtained in the densely populated suburbs of London. It is just the other end of the scale.

You enquire about Bill 2, mother. Alas! He has gone to a fate unknown. He strayed one day – jumped out of the car, I am told, in Sekondi and disappeared. This is rather strange since his father disappeared in somewhat similar circumstances. The natives here eat dogs, 'akwai mai dayawa' – there is oil fat plenty they say. It was a blow to me when I landed. Now I am 'dog less' and was very lost at first. I still sometimes imagine a dog is with me. It is hopeless country to get fond of anything – except perhaps oneself, and that is a bad thing.

My present trek has been a pleasant one, and I expect to be back in Zuarungu tomorrow. Much of it is though new country to me, and along bush routes I have not previously followed. I need to see if a big area of bush, which I have tentatively selected for a permanent rotation experiment, is typical. I have been up before sunrise, breakfasted and off by 5.30 am every morning and gone ahead of my staff and carriers to try and shoot some bush meat. We came across plenty of fresh spoor everywhere but didn't get a single shot, due I think to:- a) Much of the grass not being burnt yet, or else burnt too early and badly burnt as a result; b) The wooded nature of the bush, resulting in the beasts seeing us before them and making off; c) The bright moon in the early hours when the stuff moves and grazes.

On Saturday night, a hartebeest just put his head over the walls of the Rest House Compound and snorted, awakening my boys who were sleeping outside on the ground. But as I say I have enjoyed my trek –

sleeping out every night, there is still a touch of the harmattan about. This unfortunately obscured much of a magnificent view from Kurugu Rest House– surely the best aspect of any in the district.

I am high upon a rocky scarp, overlooking the valley of the White Volta and the southern side of the Tong Hills, 10 miles away. I have a magnificent view of the northern aspect of these same hills from my bungalow in Zuarungu – seven miles away. I forgot to mention that we came across hippo tracks near the river before crossing. The headman begged me to shoot one which had been annoying the women collecting water, and the men who made tobacco farms on the banks of the receding water in the dry season. The Kambonaaba (92) said the beast would pop his head up in answer to a call – but it didn't. I had no time to waste so pushed on. I was not sorry because I think something bigger than a .303 should be used for hippo.

And now to bath and some dinner it is already late, but fortunately at this season reasonably free from mosquitoes…."

The serenity of Zuarungu was again upset in February 1934, when Buck McGuiness and Jock Reid started up a gold mine at Nangodi, (93) 12 miles up the road towards Bawku. Apparently, the pair of them were prospecting in the area during the previous rainy season. They had almost given up hope of finding anything and were waiting in Nangodi Rest House for transport to take them south. Reid awakened from his afternoon rest and saw some samples of quartz thrown down by his partner. He pounded and panned them while waiting for tea, and the result showed a tail of gold. With the aid of hurricane lamps they retraced their steps to Nangodi market place where the samples were obtained. A gold-bearing vein of quartz ran across the area; in fact the market women sat on lumps of quartz containing traces of visible gold. They were selling items for as little as an anini, 1/10th of a penny. A mining option was promptly taken out and a concession granted. It was interesting to see a mine develop from small beginnings. The gold seam was small and narrow but rich. A ball mill was installed with a suction gas engine to drive it. Colonel Gibbs, the District Commissioner Mamprusi, and I were deputed to indicate suitable firewood- cutting areas in the bush country towards the French frontier. McGuiness, an Australian, was a great character and fine raconteur. He was balanced by his partner Jack Reid, a sound Scot who had gone into a contracting business which got into difficulties in the depression; he came to West Africa

89

originally as an official of the Bank of British West Africa. The mine contributed a lot to life in the locality by way of a market for food and labour; it was never developed on a scale big enough to upset the local way of life, although it made a useful contribution to the local economy. Apart from wages, guinea corn stalks were purchased for fuel in the suction gas engine. There was a song in Nangodi about the King of England, being short of gold, sending a message through the Governor and Chief Commissioner to the Nangodinaba. After eight years, having worked through the tailings with a cyanide process, it closed as quietly as it had opened. Years later during the war this extraordinary pair, McGuiness and Reid, salvaged the "Sangara", which had been sunk in the Accra roads. After recovering most of the cargo they patched up the ship sufficiently for it to be towed to Lagos for repair, thereafter selling it at considerable profit.

2nd March 1934 Zuarungu "Dear Mother,

I am settled now on the plinth in my long chair with a pipe and a sundowner. (94) The world seems indeed a pleasant place. 'Tis Friday night and this week, looking back, is a good one. Things seem to have conspired for the best, and the stage is now set for the season's work, waiting only for the rain for the fun to begin. The programme has at last formulated itself in my mind, and is working out on paper and on the ground. My staff are working well, and seem to know their job, and I have no domestic palavers. Long may such a state of affairs obtain, but alas! Everything is ripe now for some mighty setback. I am not looking for it however.

Tonight, shades of Mr Wright the riding master; I was endeavouring to initiate the Fortes into the noble art of horse riding. (95)

Well, March the hottest month of the year has begun. Two days ago the wind veered completely round from the north to the southeast, the change from dry harmattan conditions to monsoon. The difference in humidity is appreciable, and I have been pouring with perspiration. Another amazing thing is the sudden spurt of growth in shrubs and trees, bursting into bud although we have had no rain since November.

The Navrongo district is much less densely populated than this, and the people appear to be better farmers, but are not such cheerful rogues as the Fra-fras. There are three tribes with different languages – Builsa, Kanjaga and Kassema.

The mail has just arrived – a weekly local, not the English. The wretched Adogo missed the lorry and has walked in to Navrongo and back today – 44 miles. It was mostly official, but a grand one from Symond, one particular paragraph is so good and typical, I quote: 'Don't get too wrapped up in the job up there, old boy, or you will break your heart over the things that cannot be done. One dreams and plans, and knows well what could be done with a fiftieth part of the money that is wasted so freely elsewhere, but it is no use. The money won't be granted and the human being is an imperfect animal. The main thing is to come away knowing that you have done your little best and nothing else matters. The effort isn't wasted. "Thou knowest not what augment thy life to thy neighbours' creed hath lent." And thy neighbour can very well be a European one too! Listen to the old man moralising! Too much Haig!'

Who could help but work for a man like that. I know what J.E.S. is driving at – he is just preparing me for the shock that I may get at any time, to be transferred. Auchinleck has not carried out his threat of Head Office yet. But I am quite prepared. I have had longer here than it is given to most blokes from our Department. It will do me no harm; indeed there is much of the working of things below that it would do me good to learn. That I will be left alone for this season, at least, seems reasonably certain…"

6th March 1934 Fwegu "Dear Folks,

I came on trek with no envelopes and must rely upon my messenger to Zuarungu bringing them to me in time – it is quite on the cards for him to bring me shoes or pipe cleaners instead of envelopes.

Trekked in here today from Vra, 16 miles on my flat feet, so I am feeling pleasantly tired. I am renewing my acquaintance with the people and district here and have been holding meetings with Chiefs and their elders along the way. Fwegu is a pleasant rest house on a little hill, about four miles from the French frontier. I was the last white man to stay here nearly two years ago. Tomorrow I finish mail and trek to Arabe, and Thursday on to Nangodi the new gold mine.

The Fortes have now moved to Tongo, six miles south of Zuarungu. I am sorry to lose my only neighbours, but will see something of them anyway in the future. He has been infusing me with ideas of anthropology, and I must say that it is a very interesting study and one which is almost inseparable from the kind of work I am doing here. Indeed, my work is a specialised study of the social system……"

21st March 1934 Kologo "Dear Folks,

Here I am 'on trek' again, but in the Ford this time, along a bush road 10 miles south of Navrongo. The amount of stuff the little bus takes by removing the back and front passenger seats, and scientific packing, is perfectly astonishing. Bed, bath, table and chair and carrier; chop box containing kitchen stuff, stores and drinks; filter and wash basin; bucket; two hurricane lamps; my petrol lamp and books; my long gun and kitbag with my stuff; a cook and an overseer. I can travel self- contained for a week in this way. The car is running well, 4,200 miles up to date and no trouble – Alawi Allah! – Not even removed a wheel yet.

The weather has been insufferably hot these last few days. Evening lightning, distant thunder and sporadic clouds all indicate a change from harmattan to local monsoon conditions. The sweat rolls off at all hours of the day and night. The extent to which one has to adjust one's mode of living to tropical conditions is taxed and tested to the contrast. Book work is difficult and impossible between 1 pm to 4 so I just sleep. However, rain, which is imminent, will bring immediate relief, and lots of field work.

I had a big meeting with the Chief and his elders here tonight, and gave them the lowdown on the farming racket. I have just returned from the Chief's yam farm – an innovation this year – where I took him in the car. He reacted like a babe never having travelled in a car in his life before! I have been trekking a lot this month, mostly on horseback or on my flat feet, and am gratified to find that we created at least some impression in the Zuarungu district two years ago.

Last Wednesday, I was at Tongo, where the Fortes are living now. I breakfasted on the top of the Tong Hills. The chief of Tongozugu,(sic) Tingoli, (96) a fine figure of a man, is an old friend of mine. He said he would select for me a breakfast site, and sent the carriers and cook on with a guide whilst I was talking. It was indeed a unique spot. A cave open on two sides, the roof consisted of a flat granite rock about 40 foot square with a fine view to the south.

I took my 'groaner' to Tongo – the Fortes' is broken – they had some lovely records, Haydn, Mozart, Bach and Schubert. My stay was not without incident, for a scorpion strolled over the plinth as we were taking coffee after dinner and then the night was disturbed by hyenas. In the morning I found a little snake curled up beside my slippers on the mat next to my camp bed. However, no ill results attended these unpleasant incidents, and at this season, unusual visitations. The Fortes are getting some rude introductions to the country which, in a way, will do them no harm. They arrived with vast book knowledge of the country at the most pleasant season of the year. Like so many newcomers (old coaster speaking) were inclined to be a little scornful about the snags of which they had heard, and suggested what was wrong with people here. (97)

Back to Navrongo on Friday morning to pick up the mail – an English one, Hurrah! And collect the goods from the Treasury; it being the end of the financial year. Mothersill is paying out earlier. Four months in, and the fifth instalment of advances repaid, and the last of £15 custom duties repaid, so I should soon see daylight.

My latest excitement is with a plough or cultivator, which Jock Stewart has lent me to carry out experiments for the Navrongo Native Administration. Two bullocks are being broken in now. I am hoping to do great things with it, for if it is a success, we can persuade some folk to take it up, and it will have a revolutionary effect upon local agriculture.

The tourist season is drawing to a close. (98) It reached a climax on Monday when the Edwards, of the United Africa Company, called in for breakfast on their way from Bawku to Wa and Jock Stewart appeared for the same meal from Pong Tamale to Kugui; to be followed later in the morning by the Strides passing from Bawku to Navrongo – Posts and Telegraphs. I welcome these opportunities of personal contact, apart from the fact that always being stationed in the North we owe a big debt of hospitality received from folk in Ashanti and the Colony, because it makes things so much easier when corresponding officially with people in Accra and Kumasi. I doubt if we will see Auchinleck up here this season, though nothing would delight me more than to make him tread some local tithes, and then he might be more appreciative of our problems. Anyway, I think we have entrenched ourselves in pretty well up here now, in the Northern Province, and too much political pressure would be brought to bear if we dropped the work now. Whether I shall be allowed to remain is another matter. But this job is not everybody's 'meat,' in spite of the expressions of envy I hear from other members of the Department. They have no conception of the conditions, and I have got a good start by knowing the people and the political department, from the Chief Commissioner downwards. I will be quite content if I am allowed to see this season out.

Well, my dears, I think I must stop. I wonder if this fine breeze and the clouds mean rain. My bed is out still and my cook has waged me a shilling against rain tonight…"

My work was proceeding very much according to plan, particularly the new survey work in Navrongo district at Wiaga, in Kanjaga tribal area, Kologo in Nankani and Chana in Kasena. This was together with the demonstration plots in Navrongo itself. This district was not as densely populated as Zuarungu, and the people on the whole had more land to farm and more food; but they were not such cheerful people as the Fra-fras. On one occasion trekking through the bush from Bolgatanga to Kologo, south of Navrongo, an aeroplane passed overhead. This was a very unusual sight in those days. (99) I commented to my steward boy, Abugari, that we should travel by air in the future. He thought for a moment then observed that a "Jirigin sama" (skyboat) – would cost a lot, better save my money and marry some fine woman, sage words indeed.

Trekking in the rains could be quite laborious and sometimes hazardous for all concerned; this is well shown by a laconic entry in the diary of my overseer in Navrongo, Yakubu Singa. "September 13th. Drowned in the Chuchiliga River, saved by Sulamani." I particularly enjoyed visits to the Chief of Chana, a fine upstanding leader of his people; as a young man he had gone down to Ashanti and worked as a carpenter– he was a handyman and good to work with.

29th April 1934 "Mother dear,

It is rather nice hacking round Zuarungu, not so much for the 'going' which is exceedingly stony, but because I know lots of faces, and many more know me. Everyone has a greeting or salute of some sort and I can cough out in Nankin 'Hozanuri' good evening; or 'Oyidigansoom' is your house well – much to their delight needless to say!

I spent the greater part of last week in Navrongo enjoying myself enormously. I got the bullocks we were breaking in some time back to work the plough. It gave me enormous pleasure to hold the handles. My overseer and the labourers thought I had gone mad I think. But it really was one of the most genuine thrills I have had for a long time, just to see the soil creaming off the mould board. I am getting one or two experiments going on the Native Administration Farm, which promise to be vastly interesting and represents an entirely new method of agricultural work for the Gold Coast. Fortunately, Captain Mothersill (100) the DC is one of the very best. I have given him pertinent agricultural literature to read, and sown fertile seed. He is as keen as I am for his district to be ahead in agricultural development. In fact, my trouble now is to curb some of his enthusiasm for development, or we will be rushed along too fast; (more dangerous than no development).

I have a mason building me a bullock shed and pen here in Zuarungu, and a bullock cart is on its way up from Tamale. The Chief of Gonja, a progressive man and a great friend of mine, is supplying me with bullocks. I hope eventually to have new ones coming in all the time to train, and then send them out into the districts again. Also I want to take on learners, to be paid at labourer's rates, especially Chiefs' and headman's sons and anyone else to be trained in animal husbandry. They are especially needed to care for working bullocks, ploughing, and the manufacture of compost manure. The next step is to make some cheap form of plough, made from local materials by local craftsmen.

What a grand time I could have if my department would only leave me in peace for a few years. I am going to angle for more staff, and try and carry out the same sort of fundamental investigation work on the western side of the Northern Province – in Lawra, Tumu and Wa districts. Then my position will be impregnable, for I will be the only authority upon agriculture in the whole of the Northern Province. I am on the best of terms with all the Political Departments, from the Chief Commissioner downwards. His Excellency looks favourably upon agricultural problems up here, so may bush ways continue to hold, and may I be allowed to grow senior, wild and bush in the NT forever!

But it does not do to bank too much on these things when one's life is at the mercy of a human chess player like Gilbert Auchinleck (who has not visited the NT now for two and a half years, which is disgraceful). Tomorrow I return again from Navrongo, going down to the White Volta River at Pwalagu, on Wednesday. Here I leave the car and meet Symond, who is going on local leave in the Tumu district. When he gets back from local leave he has to go post haste to Accra, and I have to go down to Tamale again in June for 10 days. I will sign vouchers and pay out at the end of the month. Unfortunately, the drift at Pwalagu will probably have gone at the end of May, so I cannot take the car with me, and it will mean leaving the Northern Province at a critical time, i.e. the time of harvesting early millet, but it cannot be helped. It will be a good experience for me to run the whole Agricultural Department of the Protectorate for a little while.

Next weekend I go to a political conference at Gambaga to deliberate upon 'Direct Taxation'…"

We had decided that a policy of mixed farming was correct in the Northern Province by which we meant the integration of livestock with crop production. To that end, we introduced carts, ploughs and cultivators and started to break in the local oxen for work.

This did not prove such a difficult job as expected, though we had a bit of a rodeo when we started. The trainers acquired a facility after a time, and the training of oxen became almost a matter of routine. It was a great thrill to see the soil floating off the mouldboard of the plough; something revolutionary and far more effective than the small hand hoes currently used. We realised that the plough had to be introduced with caution, because of the dangers from soil erosion. With rainfall of high intensity emphasis was placed, from the beginning, on contour farming.

Fundamental also, to a concept of mixed farming, was the making of farmyard manure; the value of compost was already well known. We advocated a conscious effort to use livestock to produce the maximum amount of manure. Work in Nigeria had shown that comparatively small quantities, up to four tons per acre, were effective in raising yields. Our experiments confirmed that this amount was optimum under our conditions too.

In May, it was arranged that I should go down to Tamale for a few days. This made a pleasant change for me. I enjoyed renewing my acquaintance with Tamale Agricultural Station, where conditions were so different from those in the North. One of the jobs that I was required to do whilst in Tamale was to prepare estimates for grassing down the proposed new Tamale airport with doob grass. (Cynodon dactylon). Whilst I was away locusts (101) threatened the North, but not in large numbers.

I was interested to meet a visitor from Nigeria about this time, John Pedder from the Department of Agriculture. He came round with Jock Stewart to explore the possibility of purchasing a small breeding herd of our short -horned cattle, which had a considerable resistance to trypanosomiasis. He was locating a source of supply which he intended to pick up on his return from leave in the UK. His idea was to trek across country from the Gold Coast through Togoland, to Ilorin in the Nigerian middle belt. People were very reluctant to sell but he eventually got sufficient to make the journey worthwhile. From him I heard of the developments in mixed farming in Nigeria with which we were already familiar, through the excellent book of Faulkner and Mackie, "West African Agriculture".

It gave me great pleasure when preparing the estimates for the Northern Territories; this included provision for my passage to the UK the following year, which cost £104 return. I had a very good clerk in Tamale, Samuel Grant Mensah, who like most of our clerical staff came from the South. He was well educated and erudite, and would draft what he regarded as suitable replies to any incoming letters. When he brought them up on the file, it was very seldom that I had occasion to alter them before they were typed. I was able to visit our substation at Yendi, an interesting place started by the Germans in Togoland, where they had established an excellent plantation of kapok trees.

I was given, by Dr Vaughan, a terrier pup "Ben" to replace the dog I lost in Bongo in 1932. A dog was more or less essential for company in the kind of life I was living. I had already acquired a large male cat named "Percy" in memory of Colonel Percy Whittall, the last Provincial Commissioner in Navrongo. He had been the cat's original owner, before Captain Mothersill took over. Percy settled down in Zuarungu and was splendid company, tolerating the new puppy and keeping me free from mice, rats, snakes and scorpions. I acquired Percy after one of the many lunches I had with Mothersill, who invariably ate a curry lunch. It was no ordinary curry either, as he added more red pepper and maintained that it was not very hot. Woe betide you if you took him at his word!

One evening when Colonel Gibbs, the District Commissioner Mamprusi, was dining with me a loaf of new bread appeared on my table. Enquiry elicited that Abugari the houseboy had made it. When Abugari went out to get the next course "Gambaga George" observed that if the boy was speaking the truth he should be my cook, so he became that, getting better and better for as long as I stayed in the Gold Coast. Anasuri Fra-fra became my steward boy, a bright and amusing young man. I found him one morning on the steps behind the house teaching himself to read English from an illustrated primer and solemnly repeating JAR – POT.

Allan Kerr (102) was posted to Zuarungu as Assistant District Commissioner in August 1934, and a strong friendship developed. Soon after arrival he was required to preside over an inquest on the battered body of an old man brought in from Zokko. None of the witnesses would agree it was a man. A woman and a girl were working in a field harvesting groundnuts when they were attacked by an animal, described as being like a low brown donkey. It gave them severe scratches over the head and shoulder, they screamed and the animal ran into some long grass. It was beaten up by men with sticks; it vomited two frogs and turned into the body brought in to the District Commissioner. It was significant that an alleged man-eating hyena was killed at Winkogo not many miles away shortly after. But an innocent Moshi man resting in the grass was beaten to death.

22nd November 1934 on trek Navrongo "Dear Folks,
It is very hot and dry now here, but cold in the early mornings. Really a most pleasant time of year. I will have finished all I can do this season in about 10 days from now, and expect soon after to descend to Tamale. It

will take me a long time to work out my data, which has accumulated to a formidable extent.

Our only excitement has not been too pleasant. Syme, the DC up in Bawku, was shot in the chest with an arrow, lodging in the lung. He was trying to arrest some chaps at night, with police, but is now off the danger list. (103) And a dear old miner man named Inansi died aged 62, who was injured a while ago by falling masonry in the mine and fractured his pelvis…" (104)

Stewart Simpson the Veterinary Officer convinced me of the possibilities of short wave radio reception so sent away to a firm called A.E.R. in England. For £8 I received an excellent small set, which worked from one cell of a car battery. This close contact with events in Britain had the effect of making me feel more appreciative of the local scene.

At the end of September, I stupidly strained a cartilage in my knee by getting up awkwardly from a chair. I hobbled down to my car and on to Pwalagu ferry to pick up Dr Hawe. He came up from Tamale at the request of the Chief Commissioner to bring adrenalin for Jack Pym, the policeman in Navrongo, who had asthma. He attended to Mothersill who was said to be sickening from black-water fever, and Allan Kerr who had a fever in Zuarungu. Dr Vaughan had gone on leave from Navrongo and there was no immediate replacement. Simpson and I were the only fit people left, and I had a stiff leg and could not ride or walk. Jim Syme, DC Bawku, was causing much anxiety as a result of an arrow wound in the chest. The medical department tended to be a law unto themselves, but they were a tower of strength in time of trouble. They made a big contribution to our wellbeing. I was surprised that Dr Hawe did not wear a hat when we went round Bolgatanga market in the middle of the morning. Even the clerk in Tamale, Samuel Grant Mensah, took the precaution of covering his head with a book when he came out of the office to give me a message.

The arrival of "tick birds", the white cattle egret, late in October marked the onset of the dry season. One could sleep out again in the open without the risk of getting wet. The drifts were built up again over the Volta River at Pwalagu and Nasia, thus opening the roads to traffic. It was a busy harvesting period on the farms, with the incoming results of the surveys and experimental plots. When they were completed I went down to Tamale in December to write up my annual report. Now we had some facts and figures it was possible to put forward a realistic programme for development.

The new Chief Commissioner, a Welshman, was familiarly known as "Kibi" Jones. (Later Sir William) (105) He was a tremendous enthusiast and a great worker expecting everybody else to be the same. He had a great effect on the morale in the Northern Territories, and produced what must have been one of the finest administrations in the Colonial Service. He was particularly good at securing mutual co- operation between the Administration and the so called "other departments". He welded all together as one service, with a common policy. Practical form was being given to Native Administration through indirect rule, along the lines of Lord Lugard's "Dual Mandate". As Allan Kerr put it to me once, our function was to give instruction not instructions. This also suited the Department of Agriculture well, because we had long been aware that it was impossible to force people to farm better if they did not want to.

A Northern Territories Conference was held in 1934 in Tamale just before Christmas, attended by all officers of the administration and departments. This provided an excellent forum for putting forward our aims and thoughts. Mixed farming, soil and water conservation and agricultural education were subjects on which we as a department held strong views. The making of farmyard manure, inherent in our mixed farming programme, led us into conflict with the Health Department as regards fly breeding. We agreed the necessity to draw a distinction between urban and rural areas. The pros and cons of grass burning were pertinent to soil and water conservation. Under education we had to dispel the idea that agriculture provided a good outlet for strong boys without much brainpower. It might have taken years of correspondence to clear the air, but in this way we were able to do it in a week's discussion.

Tamale was very full for the Conference and Christmas holiday 1934. Bill Walker, District Commissioner Krachi, was staying, with Cowan. Morris Greenwood, the Departmental Chemist, was staying with Lloyd Williams, and Ted Ellison, District Commissioner Lawra was with me. We had tennis and polo tournaments and a race meeting during the week. For the New Year holiday, 1935, we had an agricultural party at Yapei Rest House, about 30 miles west of Tamale on the Volta River. The harmattan was blowing so it was quite cold and invigorating, demanding a sweater in the day and two blankets at night. The idea was to catch fish, in this we were not successful, but we greatly enjoyed the peace of the bush. There were imposing riverine trees, Anogeissus leiocarpus and Daniellia. We were entertained by hippos that were curious about us, flapping their ears, snorting and approaching a little nearer after every dive.

Life continued happily in Tamale with the analysis of yield data for last season, and preparation of the annual report for Zuarungu and Navrongo. Things were definitely taking the shape of a clearcut policy of mixed farming, coupled with soil and water conservation, attention to water supplies and the extension of rice growing. The developing Native Authorities were to be encouraged to include an expenditure heading for agriculture, which they were very willing to do. It was considered appropriate to charge the cost of extension work to the local government. This gave possibilities for general development, with agriculture inevitably an important component. The Native Administration derived their revenue from a nominal poll tax, (106) which partly replaced the tributes paid to Chiefs, who now received salaries.

To help in giving an agricultural bias to education, demonstration farms and gardens were established attached to Government and Native Administration schools. The aim was not to teach agriculture, which anyway was a university subject, but to spread an enlightened rural environment. I was pleased some time later to be given a rough lemon fruit by a Chief in a remote part of the district, who told me his son had brought the seed and the idea from school. I am sure that this policy paid ample dividends in facilitating the introduction of ploughs, farmyard manure and contour farming. Gambaga School in South Mamprusi was one of the first to receive attention, largely because of the enthusiasm of the District Commissioner, Colonel Gibbs.

Before leaving Tamale for the North we had some splendid games of polo. On my last Sunday morning we had a mounted paper chase. Sergeant Newell of the Regiment and I started from the club at 6.30 am to lay a trail to the south and west towards the Yapei Road. We went about seven miles and finished at a tremendous gallop being chased by a field of about 30 people across the polo ground, through the town, over the police parade ground and cricket ground, through the police lines and down the Conservancy Road behind the houses on The Ridge. This was so called because the prisoners removed the night-soil buckets that way. Then we went on to the Residency at the end of The Ridge, where the Chief Commissioner was our host for shandy and breakfast.

Before Xmas, I had acquired two ponies, Naba and Samari, on their way south from French country where they were bred; they schooled well for polo and I sold them reasonably in Tamale before I went north. The cost was less than £10 each and the selling price likewise, it cost 15s a month in wages to the horse boy and 15s for a bag of guinea corn (200 lbs) – so horse keeping was not prohibitive. Before departure, I acquired an excellent cold box from "the Beetle" for £12, it was a Crossley "Icyball"; this revolutionised life in Zuarungu; it consisted of two cylinders, one containing gas which was driven through a narrow neck by a blue flame kerosene lamp, and would make ice if efficiently processed. It would hold a case of beer if need be, that was forty-eight 26 oz bottles. The kerosene-operated Electrolux fridge had just become available, but it cost £32 which was a lot of money in those days.

During late March and early April 1935, we had good soaking rains which enabled us to get well ahead with cultivation and planting. We confirmed what in fact was well known to the people, that it was essential to plant timeously. The main crops sown were early millet and guinea corn, and needed even germination and good early growth. Plantings only a few days late never caught up. I was well served by good African staff who by now were experienced and as enthusiastic as I was, in bringing about an agricultural revolution.

7th March 1935 Zuarungu "Dear Mother,

Well, we are in about the hottest month of the year (107) and, with the kind offices of the Icyball, we are surviving it pretty well. I was amused last night at dinner with Colonel and Mrs Gibbs, who are staying here for a few days. The Chief Commissioner was also up for the night, and I said that I thought the only thing to do in the afternoons between 2 pm and 4 pm was sleep, and Kibi Jones said 'I agree. The man who says he works in the afternoon is either a liar or an idiot'. Incidentally, I believe he works most afternoons. The temperature keeps to 100° or over until about 5 pm these days. On Monday the Earl of Plymouth passes through, and Kibi says I must meet him. How do you address an Earl anyway?

Well, my arrangements for these seasons' experiments are well in hand. I am starting to build myself a new garage, and my well near the bullock pens is already 16 ft. deep. It's really in no sense a bad life taking it by and large, as they say. It helps a lot when one's work is appreciated. My trouble is that the process of consolidation is not spectacular. I have said in a hundred different ways all that there is to say, until such time as our experiments bear fruit. Interest must not be allowed to flag in the meantime.

In less than three moons time, Akawai Allah, the young master, will be on the water again; looking back it has been an exceedingly pleasant tour. From a personal point of view much ground has been covered, a few more rungs have been climbed. The final point will be reached if this thesis is accepted by the academic board in Trinidad. It has not been sent off yet, through pressure of work. But I have just corrected the final transcript and the illustrations were put in last night. My great joy is that I have not been south of Tamale the whole tour. This is a great achievement in our department, where lack of continuity is the order of the day. The radio comes howling in. The small motor car continues to do its stuff with the minimum of attention.

Ben is in good form and liked by all, although today he is in disgrace. I have not spoken to him since I awoke; he left before dawn to sow wild oats and did not return until 9 am, very bedraggled. But I cannot punish him for too long. I like to see his tail going and his bright little brown eyes full of fun and mischief. It is for his own good, however, that he is tied up and in disgrace. One cannot allow him to roam as he wants to, in a country where hyenas by night and Fra-fras by day are ready to chop a small well-fed dog. Gordon Cowan is going to look after him when I go on leave.

I am off early tomorrow by a bad road to Tongo – six miles along the hills to Gorogo (108) – to see the laying out of experiments in the head man's compound farm, there and back the three miles to Tongo Rest House, for breakfast with the Fortes…"

28th March 1935 On trek at Wiaga "Dear Mother,

"I am reclining back in my usual slack way in my long chair in the compound of Wiaga Rest House, watching the sun go down and listening to a talk on Thursday Island. I have my radio with me as you will guess – just a pole up 13 ft. and a wire thrown over it for an aerial, and now it works better.

I left Zuarungu last Thursday and hope to return later this week having visited our recording compounds at Chana, Wiaga and Kologo in this district; and arranged the planting programme for the Navrongo experimental area.

Of course, last week's excitement was His Lordship, the Earl of Plymouth's (109) visit. I had quite a long chat with him. He is very interested in agriculture, and has sufficient facts to see what we are driving at. I got a word in about continuity up here, with staff interested in mixed farming. I think I staggered the old boy by saying that there are only two agricultural officers north of Kumasi. He was vehement in his demand for agricultural research, and enough staff to do it. This tested me in the presence of a number of political officers. The Chief Commissioner told Cowan in Tamale that he thought I put my case very well. I am pleased, as I wondered afterwards if I hadn't put things a bit too strongly. But I cannot mince matters. I have found that it pays, so far at any rate if one is sure of one's premises, to say what one thinks even if it is unpalatable or disturbs the beliefs of the powers that be. I have reason to suspect, however, that this attitude has not made me too popular in HQ in Accra. I have now the backing of the whole Political Department up here, from the Chief Commissioner Northern Territories downwards. I get good write-ups in our recent publications 'The Monthly Newsletter' and 'The Gold Coast Farmer. Chunks out of my reports appear every month, so evidently our journalistic Chief thinks my stuff has at least paper value.

Next week, I go to Gambaga to arrange a planting programme for the Government Station Farm, and advise upon the teaching of natural science. Note that it is advice we give. 'Advisory capacity' is a grand attitude, necessitated anyway by our shortage of staff. Well, it is all a move in the right direction. We cannot change the old farmers. The young, ideally in the schools are the material to work upon, and in my opinion should be run on the most up to date lines we know. There is a strong agricultural bias anyway in the school curriculum.

The Chief of Wiaga has been up listening with my second headphones which you so kindly sent out. His children have been listening in turn. It has given me great pleasure, after a few seconds, to see their faces totally light up with a profusion of white teeth and eyeballs…"

In the middle of April, I went on 10 days' local leave to Lawra and Wa districts. I intended to stop at Nakong, Batiasan and Tumu (110) to do a bit of shooting and study conditions in what was new country to me. At Nakong, in the extreme west of Navrongo, I took a walk in the bush in the evening towards the Sisilli River. The area was littered with the ruins of old compounds, and the only explanation for these I could think of at the time was Moshi slave raiders during the last century. But closer examination might well have shown that the real cause of depopulation was soil exhaustion, followed by trypanosomiasis as a secondary factor. It was a reasonable assumption that the earliest settlements had been along the rivers. Chidlow Vigne of the Forestry Department complained that in

Navrongo and Zuarungu districts the people lived on the watersheds, where ideally he would have liked to site Forest Reserves. A herd of elephant had passed through the area during the previous rains. The bush, in a swathe half a mile wide, was difficult to negotiate as a result of countless footprints 2 foot deep. Many quite large trees had been upended.

Next day, I crossed the river and stayed at the pleasant Batiasan Rest House, which consisted of two roundhouses joined by a covered veranda, the whole well- sited on a little hill. I was thoroughly enjoying the freedom of local leave after 15 months of continuous application.

> 14th April, 1935 Batiasan "Dear People,
>
> ...Early tomorrow morning, 'gari ye waye' – the town wakes – I am going to bush. This rising and breakfasting in darkness, so as to enter the bush just as soon as it is possible to see the path, is the sort of thing which leaves permanent impressions. It will later be a buckler against the depression of a possible office stool. The uncanny rising in that quiet half hour before dawn, the growls of the dog as the hunter arrives, the looking to rifle, cartridges and drinking water, the smell of the bush as the sun warms it and the song of the birds; the stink of the hunter and his trappings, the possibility that each silvery glade holds game of one kind or another. The absurd craning to see if some stone, tree trunk or anthill is not a monster bushmeat — it never is. If one sees meat one sees it, but in the first hour, with a little imagination, everything holds possibilities. After three hours one just plods along behind the hunter, pouring with sweat, thinking mostly of long cool draughts of beer. One almost hopes that there will be no more stalking for the day until meat is sighted and all else is forgotten..."

The next night was spent at Tumu, which used to be a district headquarters. The rest house was the District Commissioner's old house of splendid proportions; R.S. Rattray (111) had at one time lived there. He must have been an outstanding personality for, apart from being a noted anthropologist, he had been a professional big game hunter in Tanganyika. He was the first pilot to bring a plane to the Gold Coast, crash landing it in Kumasi in the late 20s. He shot an elephant soon after it had passed through the compound at Tumu, and one night a leopard took his dog from under his bed at the front door. Small wonder the place was said to be haunted. Amongst the papers in the bookcase was an account of a journey to Timbuktu, made in the middle 20s. Vaughan of the Survey Department, and

Riley, a Transport Officer, completed it on a Zenith motorcycle combination. Those were certainly the days!

I enjoyed my visit to Lawra where I stayed with Ted Ellison, the District Commissioner. "On the Sunny side of the Street" was a popular tune on his gramophone. We watched the sun go down over the vast peneplain from Birufu Rest House. I thought that somewhere near would be a good site for an agricultural station. I spent a night in Wa where Graham Ardron was the District Commissioner and returned to Navrongo via Hian. It was clear that it was desirable to post staff to the West, probably to Lawra District, when available. The population pressure was not so great there as it was in the East.

12th April 1935 Zuarungu "Dear Folk,

I have been mighty busy lately. We have had some decent rains, and every cultivation has gone ahead rapidly and, as far as I can see, without a hitch. This is thanks to my African staff who are doing an excellent job of work just now and of whom I cannot speak too highly.

Today I am ploughing the field of a local farmer as a demonstration. It has collected quite a crowd because the idea of bullocks working is quite foreign to them; also pre-sowing cultivations. Unfortunately, there is a funeral custom being held in a neighbouring compound which is proving rather a counter-attraction. This represents a first step in extension, and it proves that this is not just white man's magic when it works on his own farm, for I just sent up a couple of labourers with the plough and left them to carry on…."

The time had now come to prepare for leave; Hamish Grimm was due to relieve me. I joined the "M.V. Adda" on 2nd June 1935 in Accra. Five months leave passed happily, and with the mind and body fully revived I felt it was time to get back to a job of work. One could always tell when this point had come.

Chapter 5 - 4th Tour 1935–37

Zuarungu Experiments, Nigeria, Bulletin on North Mamprusi and Marriage

I returned at the end of October on the R.M.S. Apapa. My Director was on board, together with Mr F.A. Stockdale (later Sir Frank), the Agricultural Adviser to the Secretary of State for the Colonies (112), who was travelling with his secretary MacLean on an advisory tour of West Africa.

> 2nd December 1935 Zuarungu "Mother dear,
> It is like old times sitting on the plinth listening to the wireless, which is working very well.
> I have had Dr and Mrs Irwin staying with me for the last few days. He has a science D. Sc. and is in charge of Agriculture at Achimota College; a very interesting and clever man, with whom I have travelled a number of times. He is frightfully keen on what we have been doing here.
> Ah – the announcement says it's a bitterly cold night at home tonight. It has been infernally hot here all day – 96° at 2 pm. Now it is cool, and soon I will be creeping into my bed, which is out on the plinth. One can have such a lovely sleep on these harmattan nights…."

I much enjoyed a visit from Mr Stockdale, who came up to Zuarungu for a night in mid- December with my Director. There was not much to see on the farms at this time of the year, but the general conditions were apparent and I was much heartened by our discussions. The following week I went down to Tamale to write my annual report and to discuss the next year's programme. I spent my sixth consecutive Christmas in Tamale and enjoyed the hospitality of Eustace Rake and his charming wife Muriel; they were living in the Residency in the absence on leave of Kibi Jones. Our new Governor, Sir Arnold Hodson, (113) (Ras Hodson to some because of his previous Abyssinian connection) came up in January, and after dinner one night at the Residency shook W.R. Griffiths warmly by the hand. He said that Mr. Stockdale had spoken highly of his farm in Zuarungu, to which Dr Griffiths replied "Thank you, Sir". Sir Arnold frequently muddled people up, and we never knew whether it was deliberate. New Year was happily spent at Yendi, 60 miles to the east, with Harold Blair, (114) now the District Commissioner, and his wife. Both were musical and played a dulcitone as well as a ukulele. Harold was a Higher Standard Dagomba speaker with a great knowledge of Dagomba history and customs.

In January 1936, both Britain and the Empire mourned the passing of H.M. King George V, and I was able to listen to the proceedings on my shortwave radio set. It was an eventful month because I achieved a good pass in my Hausa Language maintenance examination, and then Gordon Cowan and I went down to play polo against Accra for the Tamale Cup. The teams were entertained to dinner at Christiansborg Castle by Sir Arnold and Lady Hodson. After the Loyal Toast cigars were passed round, but there were only two in the box; one was taken by His Excellency and the other was passed to Colonel Bamford, the Inspector General of Police. He resourcefully called for a knife and cut the remaining cigar into eight pieces for the rest of the party.

We returned via the agricultural station at Kpeve in Togoland, staying with Marden Cook. We were entertained by his wonderful imagination, his two pie dogs, Whimper and Subaltern becoming Moshi hounds, and a strip of grass like a cricket pitch we were told was for "tent-pegging". There was also a story about the Sudan Camel Corps which started "You know the noise a leopard makes going round your tent old boy". West Africa tended to accentuate idiosyncrasies; Marden was a kindly person and a splendid host.

Whilst in Accra, my Director informed me that I should visit Northern Nigeria to study their mixed farming set up. I was greatly pleased at the prospect because in these days, before air travel, there was little interchange of visits with neighbouring territories. In many cases, conditions were comparable and problems similar. My programme of work in the North, for the coming season, was approved in Accra. It promised to be an interesting year, with emphasis beginning to be placed on extension work. On my way back to Zuarungu, I spent a night with Jock Stewart at Pong Tamale.

18th February 1936 Tamale "My dears,

For the last week, I have been starting a Dagomba agricultural survey around the experimental station. It has been extremely interesting for they are really good farmers, the Dagomba – far more promising material than the Fra-fras around Zuarungu, and quite different in their farming systems.

In this absurd craze for export crops, the economy of the local farmer has been entirely forgotten. Never have we had a standard for comparison of our results on the experimental station with the farmer for whom, presumably we are catering; a most unsatisfactory state of affairs.

I expect to be pretty busy in the North on my return, and want to get ahead with things. The reply from the Nigerian Government may come along at any time, and I will have to push off for a month or six weeks at a very critical time of the year. But the experience should be extremely valuable and help to consolidate my position here, in the savannah country, where cattle can be kept and mixed farming is possible…."

29th February 1936 Zuarungu "Dear Mother,

My new loudspeaker is rendering 'In town tonight'. Really it is amazingly civilised for the 'ye bushes' (This is how Rohr, the Swiss German who keeps the U.T.C. stores in Tamale describes 'the bush').

I left Tamale last Sunday and arrived at Pong Tamale for lunch, and stayed the night there. Jock Stewart is a marvellous host. He has discovered and is utilising some amazing old water cisterns known as 'buligas'.(115) Apparently, they are common over Dagomba country, although the Dagomba know nothing about them and they are all silted up. They consist of flask- shaped chambers in the ground, under a cap of laterite rock. When clean they are 12 foot deep and up to 24 feet in diameter. It is thought that they must be at least three or four hundred years old. The local Dagomba neither made nor used them; certainly storage was the only solution to the water problem in sandstone country. Perhaps the buligas were associated with the large hidden heaps, found in parts of Dagomba, suggesting a previous population. There were reputed to be many buligas throughout Dagomba. They would, of course, have to be looked for, because they had become completely silted up and covered with vegetation. We watched the excavation of some new ones he had found, and then did a grand tour of some 12 miles along a bush track besides the river. He has been doing tsetse clearing here. We finished up at a grand lake some four miles long and half a mile wide. Jock has got one of the new Chevrolets with individual wheel springing, and it is a most comfortable car in which to travel…."

4th March 1936 Zebilla "Dear Mother,

The scene changes.

Yesterday I drove down to Bawku. Syme is the DC there, and his new wife is out with him we travelled out on the boat together. They are a pleasant couple. I dined with them last night.

I interviewed the Syrian agent here, at Bawku, and learnt some details of a ground nut export industry to Kumasi. Then inspected the proposed site for the new Bawku Native Authority's school farm, lunched with the Symes and drove on here this afternoon. Had a meeting with the Chief and his Elders, and explained my mission, and here am I having a read in my long chair waiting for dinner.

The country here is rather pleasant. It is terribly denuded of trees, but about two miles away and stretching all along from east to west there is a most attractive range of granite hills. These are very similar to the Tong Hills, which reminds me that I believe that Fortes is returning to Tongo this year. (116) He recently published a terribly highbrow paper on 'Culture Contact' in 'Africa', the magazine of his institute which is full of technical jargon. He talks about the 'magical aspects' of agriculture, (117) and asserts that Government makes no headway with her agricultural policy through ignoring these magical aspects.

Well, we have started March – the end of the financial year, and therefore a month of considerable humbug and trouble. Unspent balances in votes are repaid into the Treasury. £150 had been voted to put a tin roof on my bungalow, which I have agreed to supervise. I know nothing about it needless to say, and the end of the bungalow is rounded which does not help matters. The tin will have thatch on top. It should make a nice weatherproof roof, and the bungalow should be very much cleaner..."

March 1936 was as usual a very hot month in Zuarungu with clouds heading north up the river courses. We had a good deal of thunder and lightning but no rain. I was busy laying out demonstration farms attached to schools in Bawku, Gambaga and Sandema, plus the White Fathers at Navrongo. Their secondary school was keen to follow our advice. These farms were essentially an intensification and extension of the local farming system, the difference being in the use of local oxen to pull ploughs.

An imported iron ridging "Victory" plough (at the cost of £2 30s) was used, made by Ransoms of Ipswich and supplied through the excellent agents John Holt & Co. in Kumasi. Ridging concentrated the small amount of farmyard manure where the plant could use it.

It facilitated planting in lines which made subsequent working easier and, provided the ridges were laid out on the contour, was excellent for the conservation of soil and water. Livestock were bedded down with straw and grass to make the maximum quantity of farmyard manure. A corollary to this programme was the total control of grass burning, (118) which was possible with the high density of population. Experience had shown that, with 40 persons to the square mile or more, it was feasible to ban bush fires. With less than this a light burn early in the dry season was recommended. The Native Authorities agreed to finance local schemes which augmented considerably the expenditure by central government.

13th March 1936 on trek at Chana "Mother dear,

Picture me reclining back on my long chair, within a round mud house 24 foot in diameter set on the top of a hill amongst the rocks, in an almost ideal position. Actually, there are four such huts set within a beaten compound and low wall – my living room, chop room, kitchen and boys room. I have the headphones on and the set is working marvellously with a bit of wire strung inside the house for an aerial about 16 foot long – it's almost incredible, isn't it? The only thing disturbing the excellent reproduction is an ominous crackle as a flash of lightning descends upon us.

There is a small tornado going on outside – the first of the season and the air feels fresh and cool already. It was unbearably hot this afternoon; I lay on my bed wrapped in a towel and just poured with sweat, saturating the bed and pillow. At the foot of the hill the little car sits in her garage with the trailer, having brought myself to Chana this morning. This is enabling me to have a stay of five days in supreme comfort, with two chairs and two tables not to mention all the other necessities – books, typewriter, bed, bath, radio, drinks, filter, cook box, steward boy and cook.

I have heard that I am not going to Nigeria till September, which suits me. Also, it gives me something to look forward to, and will break up my tour very nicely. I know quite a lot of the folk in Nigeria one way or another, either travelling with them or as friends at different times, and I have time to get in touch.

I shall be pleased when I have got the next month over, and all my planting finished and experiments in order. I can do nothing till the rains come, and then everything has to be done at once.

Later – Just consumed a breakfast and Ovaltine and broke out into a muck sweat. It's a tricky time of the year just now. Very hot and thundery, and scudding clouds at night make it inadvisable to sleep out. There is nothing more loathsome than being wakened in the middle of the night and having to take your bed to pieces, to get it through the rest house door. I always make my boys sleep out if I do– they will sleep through more rain than it takes to awaken me. But it's like a furnace inside the roundhouse, and its unpleasant awakening in the morning with beads of sweat on the brow.

However, today is Sunday and as I've often noticed before, Sunday is always cooler than any other day of the week. So I slopped around all day catching up on four weeks' papers, and typing the odd table for the Appendix to my annual report.

Tomorrow, I go down to Sandema, some 20 miles away where the DC is building a new Native Administration school. I am going to advise and choose a site for the school farm. There is considerable agricultural bias given in the curriculum in these schools, and a farm run as a demonstration unit. It is used as well to augment the pupils' food supply, and is a very promising channel for the extension of the results of our experiments. Our only hope lies in the next generation; we cannot change the present one that is certain, except in isolated cases, of course…"

I was pleased with the general appearance of the Government station at Zuarungu, the maintenance of which had become my responsibility. This took the place of the desultory groundnut cultivation by the Chief, which never produced much produce and encouraged erosion. Grass paddocks were demarcated for the Government cattle from the Farm Centre. It was interesting to see how the stones on the surface disappeared, and the whole places became cool and clean throughout the year. Tree growth improved and the station water supply, a series of open wells, improved also. Tip the balance of nature in favour of conservation and it was astonishing how quickly favourable results were visible. It was felt that what was achieved in a small way on the Government station could be achieved throughout the district, when the people were ready for it.

2nd April 1936 Zuarungu "Dear Mother,
Very unexpectedly, we had a tornado last night, which gave us the first rainfall of the season, a decent 1.03" which transformed the 'soil' from concrete to wet granulated sugar. Many of the people have been planting all day, on last year's stubble with no preliminary cultivation at all. I have been busy with the plough since six this morning, with a break from 12

noon to 3.30. We have been ploughing a couple of cultivation experiments in the farms of local people, thus accomplishing demonstration and experimentation at one and the same time. The layout is a series of strips one twentieth of an acre 33 ft x 66 ft five times replicated as follows;-

A native uncultivated B ploughed A B B A A B B A A B (119)

At harvest, adjacent plots are compared in relation to their own means, and the significance of the yield mean differences tested. This is the method I have more or less adopted as standard in all my trials now. The great advantage is that you are getting your plots unevenly distributed i.e. there are as many a's on the right of b's as on the left – they would all be on the right if you went ABAB etc. Unfortunately, this layout does not conform to the statistical requirements of randomisation, and the formulae which I use in calculating the standard errors are based upon the normal curve of errors – implying a random distribution. On the other hand, with my staff the layout is simpler to follow, and it has greater demonstration value. Also, I maintain that you might arrive at such a layout as I have adopted by random sampling.

Did I tell you I am not going to Northern Nigeria until September, as it is not possible to arrange a relief for me before then. Actually, I believe I could leave everything safely in the hands of my head overseer, Mr Yenli, who has worked with me for four years now and is an excellent official in every way.

I have heard from Andrea, he is in the P.W.D. at Zaria, which is only eight miles from Samaru, the Northern Headquarters of the Department of Agriculture. He suggests my spending a couple of days with him, which will be pleasant. Furthermore, he suggests my joining him in an overland trip home across the desert in his car, through Kano, Zinder, Agades, Hogger Mountains, and eventually into Algeria. It appeals to me considerably as I am sick of the monotony of Elder Dempster's luxurious liners. The snag is that the roads are bad between Kano and Agades early in May, and I shall have to get permission to shorten my tour.

I am reading with considerable enjoyment Lady Lugard's 'Tropical Dependency', which is all about the history of the Sudan, North Africa and Northern Nigeria. There is naturally much about Timbuktu. Unfortunately, I doubt if my schemes for local leave to that place will come off with this trip to Nigeria…"

The gold mine at Nangodi (120) now employed eight Europeans and produced £60–£100 worth of gold a day. It was surprising how easily it fitted into the local scene. It gave local employment, even Dumdum, a deaf mute, found a job watching the tables for gold nuggets. The local ladies found a market for guinea corn stalks, which were used as fuel for the suction gas engine which drove the ball mill. I enjoyed spending a night at the mine with Jock Reid, the Manager, on my way into Kusasi district. I left my car at the mine, and went down in the mine lorry to the Red Volta River where the road drift was washed away. We crossed by dugout canoe with the usual shouting and good humour, and carriers from Tili were waiting to take my loads on to Zebilla Rest House 12 miles away. The following day I cycled with the messenger 20 miles into Bawku. My main purpose was to discuss the groundnut trade with the Syrian traders and to inspect the school farm, which was in the capable hands of Cockra, the Headmaster, and a keen agriculturist.

Dr Denoon, a Canadian from Alberta, and his wife were the only Europeans in the station. The District Commissioner Jim Syme and his wife were out on trek. I enjoyed the hospitality of the Denoons for lunch. Unfortunately, shortly afterwards Dr Denoon died of typhoid fever as a result of drinking unboiled village water, when on trek. (121) A pushbike was invaluable in the rains; on the way back to Zebilla I nearly ran into a storm drain whilst admiring the flight of a dozen spur-wing geese.

29th April 1936 Zuarungu "Mother dear,

I seem to be continually on trek, getting back only yesterday from Zebilla and Bawku. The little car pulled through three broken drifts, and any amount of soft road in amazing fashion, where a wider and heavier car would definitely not have got through. Tomorrow to Navrongo to get the pay; then back again on Monday and to Chana returning on Friday. On Saturday, I go to Gambaga if I can get through the roads, to see the school farm. It will then be time to get back to Zebilla, and it will mean riding up since the roads have gone. I want to do an extensive trek in Bawku District lasting about a fortnight. Pushing on each day 10 or 20 miles is the way to see the country. In Bawku, I purchased a horse for myself, a great upstanding strawberry roan; well up to my weight. He should make a good trekking horse…"

13th May 1936 Gambaga "Dear mother,

I left Zuarungu on Monday morning, and did two mighty days trekking over the White Volta River and the escarpment to Gambaga, arriving on Tuesday. The trouble is that the man in charge of the farms is a school teacher first, and a farmer second. His only adviser can be Colonel Gibbs the DC who also is not a farmer. At least they are learning that farming is not quite such a simple business, as they might once have thought it to be. The show is interesting as it is the first school farm, of the three under my care, to use bullocks and a plough for cultivation, and to practise mixed farming.

I am sitting waiting for lunch with lumps of ice clinking in my glass. Such a luxury! The Gibbs are very kindly folk, and sent down yesterday a big supply in a large thermos from their Electrolux. Tomorrow, I take a holiday and go out with Gibbs to see the sleeping sickness camp at Nakpanduri, and the tsetse fly clearings. Nakpanduri is in German Togoland, and I have not been there since October 1930 when on locust control in the area.

On Saturday, I trek off again over the scarp to the North, to do a gentle tour round the Kusasi area which I am starting to survey gradually this season. I rode Marman, the new roan, to the river at Bepala on Tuesday, and have sent him round to Zongoiri to meet me on Saturday. Then I go to Kugsia for a night, and up to Zebilla eventually arriving back in Zuarungu, through Nangodi, next Wednesday or Thursday.

It is cool at this season and everything is growing and is the freshest of green. I seem to be busy from morning to night. As Anasuri said when I asked him when he was going to teach me his language – Nankaim – 'You too busy Sunday and weekday all be the same thing'. And so it is.

The trouble is that my stuff is so scattered, and things change so rapidly at this season of the year. No sooner have I returned from seeing things in one place, than I begin to wonder how things are going someplace else. There is a great kick to be had out of planting seed, going away, to return in a couple of weeks or so to see how it is getting on. But sometimes I don't get round for a whole month and much can happen in that time.

But all this makes for contentment in this country because it means interest. My household knows the racket now too, which simplifies trekking and life generally. They really are working very well just now and I seldom have to have a word with them…"

Back in Zuarungu, I sent horses to Tongo so that the Fortes could come in for the day. I always found their company stimulating; in the evening I drove them in my car through Bolgatanga to Shia, from where they had only a two- mile walk along the foot of the Tong Hills to Tongo Rest House.

7th August 1936 Zuarungu "Dear Mother,
There is no cookery book that exists which could tell an African how to cook under the conditions that these chaps have to perform. The wonder is, not what they do not produce, but what they do. When relieving my

feelings to Fortes, he paid me a subtle compliment by saying that if we were interviewing a cook, and heard that he was Lynn's cook, there would be no argument. But he's a flatterer, and has had many meals in my house.

Life in Zuarungu is nearly perfect. There can be few such places where I could have found such a joy of living. I am able to indulge any tests in a small way, as it is almost impossible to do at home, unless one is very, very lucky. My joy comes from my funny old mud house and my garden, and many books, and Ben and Marman. My latest acquisition is a shower bath, and it is a perfect joy. It consists of a galvanised iron bucket holding about four gallons, with a rose secured into the bottom and a cord to start and stop the flow. Earlier in the year, I built a grand little bathroom into the house, with a concrete floor. I can't think why I didn't think of getting one of these things sooner.

The garden is looking well, as is the residential area generally. A number of my trees are now growing up, and others I had to cut down – thinning out an avenue of Flamboyants to advantage. The large areas I have put down to grass, and my 'non-burning' campaign and no tree cutting by unauthorised persons are gradually taking effect. Before, it was nobody's job to look after trees. But it is a hobby of mine, and I have certain facilities at my disposal in the shape of a nursery, labour, tools etc. and a sort of quasi- authority where these things are concerned. Naturally, it takes years before one can see a return for one's labours, but already I can see it. I treat the whole area as my park. How much would these privileges be worth at home? The vegetable garden, too, is providing a very necessary quota to the general scheme of things – tomatoes, lettuces, radishes, cucumbers, beans and cabbages; (122) now for the first time cauliflowers grown from some Indian seed I got from Suttons. Yes, and carrots, Jerusalem artichokes, guavas, Cape gooseberries, and my first pineapple a day or so ago.

Now we have breathing space, having finished harvesting the early millet and worked out the statistics. Now the late crops, guinea corn, late millet, rice, groundnuts, sweet potatoes etc. are growing like fire, but so are the weeds! Thanks, however, to the bullocks and straight planting, the work on our own fields has been greatly simplified this season. This gives the labour time to implement my tree- cutting program, gravelling the drive and so on.

The work is at an interesting stage just at present. The season has been a favourable one so far, anyway, and we have a good grip on the situation. So it is with perfect confidence that I leave matters in the hands of Messrs Yenli and Wala until my return in the middle of October. Next year I want to put together the results of our individual experiments, in the form of 'unit holdings'; that is, the area we consider a family and a pair of working bullocks with a plough can work and maintain fertility. This area will be something like 10 or 12 acres, as compared to the two to three acres farmed at present, and the yields will be higher per acre. It is an attractive dream…"

It was now time for me to set off for Nigeria and I joined the R.M.S. Adda on the 26th August 1936. I took with me my steward boy, Anasuri Fra-Fra, who was wildly excited by the prospect. I also took my camp kit and saddlery. As I was deposited on deck by the mammy chair from the surf boat I met two members of the Agricola Club, (123) C.G. Webster, on transfer from Burma to Nigeria and de St.Croix, who was with the Veterinary Department in Nigeria and was an expert on the Fulani language and culture. The following day we arrived in Lagos harbour and tied up at Apapa wharf.

30th August 1936 Zaria NIGERIA "Dear Mother,

Then out to the Customs sheds at Apapa, which is across the river from Lagos, and on the mainland where the boat train starts. The entrance to Lagos harbour was very attractive. The waterway runs in along a wide street with imposing buildings upon it – Government House, the Chief Secretary's Lodge etc. It is much greener than Accra, and a very good harbour. Lagos is really an island, and is an enormous place. De St. Croix and I had a pleasant dinner in the Railway Restaurant at Apapa. Then, as the hour of 10 pm drew near signs of tremendous activity took place. A number of cars appeared from Lagos. Each carriage of the boat train is labelled 'Kano' – 'Zaria'–'Jos'– or 'Kaduna'. Romantic names.

The train is extremely comfortable, two in a compartment. The long seats come out a bit and make excellent beds. There is a fan, and cunningly arranged reading lamps in each compartment, and I have seldom had such a pleasant train journey. Hand basin and running water in each compartment. The restaurant car was very good, a bit expensive possibly, (1s 9d for a bottle of beer) but first class.

Well, the 'Sultan of Sokoto' pulled us through the night, through Abeokuta and Ibadan, and we awakened in savannah country in the morning. We were soon in Ilorin. I was surprised to get into open country so quickly. Ilorin to Jebba and my first view of the Niger. The river was very swollen and enormous. There is a fine railway bridge across, and I took some snaps which I hope will come out. Then another night and we awoke still in the same sort of country, very similar to that around Tamale and Navrongo. Kaduna at 6.30 am and into Zaria at 8.40 am.

I am staying with Andrea for the weekend and going out to Samaru tomorrow. Mackie, the Agricultural Director, called in to tea yesterday.

They really are charming people. Mackie is a very popular director. He suggested my itinerary should be amended to include a visit to Ibadan, the Headquarters of the Department and not confined to the 'mixed farming' country as proposed. My itinerary has all been worked out and includes stays at Samaru, Dandawa, (the Empire Cotton Growing Corporation Station) Funtua and Gusau and up to Kano by train. In this way, I shall have seen more of Nigeria than many people in Nigeria.

Andrea has a most amazing handmade gramophone, one of these super things with a colossal horn 8 ft high. The reproduction is like nothing I have ever heard before. We are listening to one of Beethoven's Quartets. A lovely thing…"

The "Sultan of Sokoto" was the impressive steam engine which pulled the boat train 600 miles up country. Nigeria was very impressive and surpassed my high expectations; I was particularly struck by the enthusiasm of everybody for what they were doing. Samaru was the Headquarters of the Department of Agriculture in the North, some eight miles out of Zaria. I stayed in the rest house and found that cream from the dairy at 1½d for a jam jar full was too cheap for my digestion! From Samaru I went up by train to Funtua, and from there 80 miles by an excellent Native Administration kitcar to the farm centre at Kafinsoli. This was an important centre for mixed farming in Katsina Province.

I was privileged to be given an interview by the Emir; he was dignified, wise and very interested in farming. He insisted on presenting me with a sheaf of "gero", the local early maturing millet, which had a head like a sword 3 ft long, whereas ours was only 8". (I planted it in Zuarungu the next year and the people were very impressed. It did not give any higher yield per acre than our own, and in a year or two disappeared entirely through cross pollination.)

I had compiled a list of 60 mixed farming questions before my visit, to ensure I covered the field. I found my hosts were anxious to supply the answers and much more besides. From Kafinsoli I journeyed by lorry to the Empire Cotton Growing Corporation's seed farm at Dandawa, where mixed farming principles were applied over the 600 acres of the station. The soil conservation works were particularly impressive; this was a matter of which we were only beginning to become aware in the Gold Coast. Whilst there we had a storm with hailstones the size of golf balls; so I was very proud at Shemi Station next day, while waiting for a train to Gusau, when

Anasuri asked if I would like ice in a drink. He had saved hailstones in a thermos flask from the previous day.

From Gusau I went south to Zaria, then on to Zonkwa via Kaduna Junction arriving at 4 am in a Scotch mist with hurricane lamps and the light from the train. The agricultural station was eight miles from the railway, set amongst attractive green hills on the edge of the Jos plateau. It was high and the air was cool. The people around, Katabs, were pagan, somewhat similar to the Fra-fra in Zuarungu, but the population density was not nearly so great. The women wore a bunch of leaves in front and a mushroom- shaped affair of plaited grass at the back, the men wore beads and plaited their hair. Their round mud houses had tall thatched roofs which came down very low at the eaves. The system of farming was highly developed; shallow ridges were pushed up with an elaborate hoe like a hand plough. It had a long narrow iron blade on the end of an iron spike, onto which a wooden handle was fitted at an acute angle. Ginger had been developed in this area as a cash crop. We returned to Zaria by train.

My next trip was to the farm centre at Maigena near Zaria, which was being run very successfully by Mr Vigo from Jamaica. He was astonished that a visitor from the Gold Coast could talk in Hausa to the local farmers; I think he was under the impression that I had learned it since I arrived.

19th September 1936 Maigena Farm Centre, Zaria "Dear Folks,

Well, my tour continues to be full of value and interest, and I feel quite at home. Maigena is an interesting place started in 1931, when three farmers were set up with a pair of bullocks and a plough. (124) And now there are 250 farmers set up, and more on the lists as fast as cattle can be trained. It is the same story everywhere, and the scheme is most attractive. 'Setting up' consists of the Native Administration giving an advance, usually of about £8 in the form of stock and implements. This sum is repaid plus 5% over a period of four years. But the railway makes groundnuts and cotton profitable crops, especially in Southern Zaria where I was last weekend. Our position is not quite so simple; the only export crop we can see at the moment is livestock at first, I think. However, a return will come from the sale of foodstuffs locally. This will give us a breathing space and enable us to develop the possibilities of livestock. We have an advantage over Nigeria in that the people already have cattle, and an advance will be necessary only for a plough. Also, I think we can charge 10% in the beginning, to deliberately slow up expansion and to encourage rapid repayments. I am very eager to get back and make a start now.

I will be very sorry to say goodbye to the friends in Samaru. I trek in from there, 18 miles on Monday. Then dine with Andrea on Tuesday night and catch the 12.30 train for Kano. The following Saturday I pass down on the 'Rich Mixed' Lagos train to Ilorin. The train stops in at Zaria station for nearly an hour, and I hope a party will come in from Samaru and have drinks with me then.

Anasuri, by the way, is enjoying this trip. He has quite settled down to things and proposed to send his 'big' brother to me to learn farming on my return. This is a very big step forward. He might easily be our first mixed farmer. Ah, I must tell you, Anasuri came in very excitedly and told me he had seen some sort of thing he had never seen before, in the front of the Emir of Katsina's Palace. It was a thing with four legs, bigger than a horse, with a long neck and small head and no horns. It had 'feathers like a sheep'. Quite obviously he had seen a camel. But what a lovely description 'feathers like a sheep!' I think he thought it was a four- legged ostrich at first. Anasuri is seeing many strange things. He made a profound discovery at the outset, when our surf-boat got swamped and the sea tasted salty. What tales he will have to tell when he gets home…"

By now I was well advanced in my tour and itching to return to put my new knowledge into practice. But I still had Kano Emirate to visit and, on my way south, Ilorin and Ibadan. Captain B. Sharwood Smith (later Sir Bryan) (125) was the District Commissioner Zaria, and I enjoyed a Sunday supper in his town rest house. It was an enormous roundhouse with a high thatched roof. He impressed upon me the significance of mixed farming from an administrative point of view. There was a short course on agriculture, for administrative officers, held annually at Samaru.

Kano was all I had expected and indeed much more. The day began with my arrival by train at 5.45 am and a visit to the veterinary establishment before breakfast. In the evening, I went up to the polo ground which was of hard sand and very fast and true. I played in four of the six chukkas and was well mounted through the kindness of Raymond Barrow of the Royal West Africa Frontier Force. The exercise was excellent after sitting about so much in trains and cars. I don't think I had ever seen so many polo players together before, or so many ponies. Unlike the Gold Coast where all horses were imported from French country, they were bred in and around Kano. As we returned from the polo ground I saw one of the first aeroplanes come in carrying mail on a regular service. (126) Amazingly, I received a letter posted in London five days before. We had little thought at the time of the implications of the development of air travel. As I left, Kano was shrouded in a harmattan haze, at least a month earlier than we would have expected it in Zuarungu.

27th September 1936 Ilorin "Dear Mother,

I left Kano yesterday and am, alas, now proceeding south. Indeed, I am 457 miles south of Kano already, but still another 200 from the sea. I arrived at the unlikely hour of 5.45 am and got right down to it, with a visit to the Veterinary Establishment before breakfast. This was last Wednesday. In the evening, polo. (127) There are plenty of playing members and stacks of good ponies. I did not let my end down badly I think, but could scarcely walk for stiffness the next day.

Thursday, mixed farming, and in the evening we drove out 30 miles to a place called Damberta. On Friday, we went north about 60 miles (by road) to Dama which is very near the French frontier on the road to Zinder. Here we breakfasted with Tallantyne, an agricultural officer in his first tour who is running farm centre there. Had an interview with the Emir of Dama, a magnificent, jolly old man who actually I had met previously in Katsina. Then we hurried back to Kano in time for polo.

Last Friday, I had an interview with the Emir of Kano in his palace. It was all rather like the Arabian nights. Kano is really a big man and withal wealthy. The audience with the haughty and dignified Emir in a vast room in his palace was an experience to be remembered, our interview being conducted through the 'Wakilin Gona' (Minister of Agriculture). The new Native Administration Offices are really magnificent. They cost £20,000 and were opened by the Governor, Sir Bernard Bourdillion, earlier this year. Then a short look around the Native town within the old city wall, and it was time to catch the train at 11.40 am which brought me to here in Ilorin this evening.

Kano is a romantic city and I would like to have had more time to see and to know it. The Hausas in their flowing robes are very attractive, and, of course, many of them are extremely cultured. Their industries are fine, that leather writing case coming from Kano. They have a written script and considerable skills in leather, iron, silver, brass, basket work and pottery.

Ward is the agricultural man in Zaria but I have not met him yet. He is out of town and expected back today. But a clerk and lorry met me at the station, and a nice little rest house was waiting prepared. Sanu!

Later. I am enjoying myself here. Learning a lot here too, as this is a different district. Met the Emir yesterday and many people including the Resident at the schools sports, drinks with the Resident afterwards and a game of polo fixed in my honour for this evening, to which I am looking forward very much. Tomorrow I catch the 4.45 pm train for Ibadan. The Director of Agriculture, Captain Mackie, (128) has asked me to stay with him when I go down, and he is meeting the train.

Down on the train on Friday to Lagos and I sail on the 'Abosso' on Saturday. What a full five weeks it has been and how I shall think back with pleasure upon the kindness and hospitality I have everywhere received."

I travelled south to Ilorin, in the so-called "Middle Belt" sandstone country, similar to much of the Northern Territories of the Gold Coast. I saw the short- horned Gold Coast cattle trekked overland by Pedder three years ago looking well and standing up to trypanosomiasis infection. Then my return to Accra on the R.M.S. Abosso (this was the newest ship of the Elder Dempster fleet).

The trip had gone well. I realised we enjoyed an important advantage over Nigeria, in that potential work cattle were already owned by our farmers. In Nigeria, they were owned by nomadic Fulani and not by the Hausa cultivators. On the other hand, the railways went deep into the interior savannah in Nigeria, providing economic transport for such cash crops as groundnuts and cotton. When the local market for food crops was saturated, we had the prospect of turning to livestock products for our cash reward.

I had to pay the inevitable price for my gallivanting, a "Report on a Visit to Northern Nigeria to Study Mixed Farming" which was published as Bulletin No. 33 of the Gold Coast Department of Agriculture in 1937.

12th October 1936 "Dear Mother,

My colleague and great friend Gordon Cowan should have sailed back on the last boat and landed on the 10th. Poor old man has had a month's extension due to gastritis, and is subsisting upon milk and boiled fish instead of beer and steak. Alas! He is getting rather tired of this country. This tour will be a 'test tour'. He is thinking of buying an island in Loch Lomond called Idimorin, 15 minutes by motor boat and two and a half miles from Balloch. However, the land owner was the General Manager of the India Tyre Co., and he invested £40,000 in it, and went bankrupt. What old Gordon thinks he will do with it, or how much capital he has to play around with, I cannot think? At least he should not be in this country. Nobody who can afford to live at home stays here.

Now I must stop. Tomorrow is a busy day with a meeting at Bongo, seven miles to the north, and one at Nangodi, 12 miles east in the evening and dinner at the mine; Wednesday Bawku, 45 miles east, with meetings in various places, then Sunday Navrongo, then Sandema and Chana.

It is a great life. Let's hope we don't weaken…"

Mixed farming was becoming something of a shibboleth in the Gold Coast. Fortunately, we had enjoyed a few quiet years in survey, so it was about time to explain our views. The Chief Commissioner was starting his annual tour of the protectorate. I was invited to travel with him and to speak at various centres in North Mamprusi, Bolgatanga, Bongo and Nangodi in Zuarungu, Bawku and Garu in Kusasi, and Sandema, Wiaga, Chana and Nakuru in Navrongo. The advantages of mixed farming, the dangers of soil erosion and the importance of controlling grass burning formed the burden of the argument. Such meetings demonstrated that Government was speaking with one voice. This was important for the European and African staff as well as for the Native Authorities and the people. It was important to do this with taxation (129) so newly introduced and misunderstood.

Locusts (130) came back in October to once again ravage the countryside near the Northern border.

12th November 1936 Navrongo "Dear Mother,

It is too hot to sleep this afternoon – I am sitting up in bed with only a towel round my waist, but it and the sheet are like the top of an oven. It is taking the season a long time to change over. However, it was nice and cool last August, and I cannot have it all ways. I am staying with Simpson (131) the Vet, but return tomorrow to Zuarungu.

His Honour the Chief Commissioner went on this morning to Tumu. Since last Sunday week I have addressed the multitudes at Bolgatanga, Bongo, and Nangodi in Zuarungu district; Bawku and Garu in Bawku District, and Navrongo, Sandema, Wiaga, Kadema, Kanjaga, Sinyensi, Chuchilaga, Chana, Keteo and Nakong in this district. What North Mamprusi does not know now about 'higher farming'; the evils of grass burning and the perils of locusts is not worth knowing! What put it into the Chief Commissioner's head to ask me to accompany him I don't know. It may have done some good. It has done me personally some good anyway. He is going to see that I get a car allowance allocated to me, and has established the principle of spending Native Administration money upon agriculture up line. But this publicity cannot go on for ever. We must stop somewhere to do a little consolidation work, so back to the Bulletin on 'mixed farming'. I am keeping very fit in spite of all these upsets and the heat. But time just slips away like a treadmill under my feet. Ben is in rare form these days but like me feels the heat. However, he can sleep all day, I cannot sleep at night.

I shed my trailer a week ago on my way down from Bawku, and since it was dark I don't know what happened to it – a wheel bearing seized up I expect. (132) Haven't had time to go back to look at it yet. Fortunately, Simpson took down most of my loads that afternoon in his lorry. Otherwise the little car is in grand trim. She has now done 23,000 miles.

Cowan should have landed yesterday if he came on the last boat. I am looking forward to seeing him again…"

27th November 1936 Zuarungu "Dear Mother,

I am still as busy as ever sorting out the results of this year's experiments. I have got the results, but the form is not coming as easily as it should, and I'm afraid my Bulletin is coming along slowly. Then I have a big building programme on hand, and all sorts of exciting things up my sleeve for next year to be prepared.

Cowan is back in Tamale, and may be coming up tomorrow to spend the weekend with me. I hope he does. I hear that he is as thin as a rake and looks terribly ill. He has just been cheated out of a lot of money by a solicitor, who is now behind bars for embezzling £20,000 upon 17 charges; so bang goes Gordon's ideas of settling down to farm at home. I'm not sure if it is not less worrying to have no money to lose.

But my bad news is that poor little Ben died last Tuesday week. (133) The best friend man ever had, two and a half years R.I.P. Simpson and I thought at first he had something stuck in his throat; then we suspected rabies. A post-mortem revealed a worm extruding from a cyst into the oesophagus. I still can't believe that he will not come snuffling up at any moment.

I am getting a lot of pleasure out of Marman however. He is in splendid trim and will make a very good polo pony I think. My latest development is a paddock between my house and the vegetable garden, where we do figures of eight every evening, and I have rigged up a little jump.

The wireless is coming through very well these days, and is a boon we take now for granted. The weather is still very trying. We are not having the usual harmattan of the season and half an inch of rain fell a week ago. Quite out of order.

I shall probably be going down to Tamale in the middle of December as is my wont. I expect to be down for about a month and hope to play in the polo game against Accra early in January – if they ask me.

Later – Sorry I missed a mail. I've been in a flat spin everywhere and am up to my very eyes in work. Things are developing; I have now money to start building my office and stores. More money to pay compensation for another 24 acres of land I want to acquire for the Government, and a teacher's course in 'mixed farming' to arrange…"

In December, I went down to Tamale to write my annual report, to attend meetings and discuss the next year's programme of work. Much time was devoted to agricultural affairs at the Northern Territories Officers' Conference; I was required to deliver a paper on mixed farming and the part which Native Administrations should play in its development. A paper by Jock Stewart, Veterinary Department, on a cattle breeding scheme led to quite acrimonious discussion. It assumed the development of a ranching approach to cattle, (134) which was the antithesis of what the agriculturists held to be needed. The Health Department represented by Dr David Lennox was critical of the farmyard manure aspect of mixed farming, from the point of view of fly breeding. The agriculturists maintained that if adequate bedding was used the heat of fermentation would kill fly larvae, and the result would be more sanitary than before. There was no doubt that such conferences saved years of misunderstanding and endless correspondence. They went a long way, influenced always by the personality of the Chief Commissioner, Kibi Jones, (135) to weld the Northern Territories into an effective administration.

6th December 1936 Tamale "Dear Mother,

Here I am amongst the flesh pots and enjoying life to the full. I have been asked to play in the polo match against Accra in January, and I am practising hard. A game of polo last night did me the power of good. I had been very depressed in Zuarungu before I came down about things in general. All my forebodings have been dispelled since arrival.

This morning I had an interview with the Chief Commissioner, upon the subject of a livestock policy in this country. This interview proved of great value, for I learnt many things about the political side, and was able to record my point of view. Up to now animal husbandry has been solely the function of the Department of Animal Health. Recently they sent a despatch home for £10,000 from the Colonial Development Fund, for a perfectly outrageous scheme for cattle improvement. It has been criticised by Stockdale, and the Chief Commissioner begins to realise he has been badly advised. We felt strongly on the matter, but our Director is apathetic.

It is all a tangle, but I pointed out that it goes against the grain to see things going wrong, in spite of the fact that the easiest course is to do nothing and draw one's salary just the same. It is impossible to become a cow in a field with high hedges and keep sane. Furthermore, the scheme is incompatible with Government's mixed farming policy. In spite of my liking for a quiet life, I am being dragged into a higher position. Heaven knows where it will finish. I firmly believe, however, if one is honest with oneself, one knows one's own mind. If one sticks with it, the point will eventually go home and win no matter who is against one at first, including the Governor. The Chief Commissioner is keen that I should start a pedigree stock farm in Zuarungu. Regretfully, I had to tell him that from a departmental point of view, Tamale was the place for such a farm.

I am very busy, but very fit, and enjoying many quiet talks with Gordon, who is one of the finest men ever to draw breath. This is a viewpoint shared by all who know him. Many officers are on trek or leave, so it will be a very pleasant stay of six weeks or so, and then an easy canter home..."

29th December 1936 at Tamale "Dear Mother,

This has been an exacting time, but I have the consolation of looking back upon it with no regrets. It is a long story with deep ramifications which I do not propose to go into fully.

It has been a good Christmas. This is a family festival, and Tamale has been like one big family. For the New Year Cowan, the Beetle, Blair and I are going to Dagomba to fish, read books and eat the Xmas pudding.

The Political Conference was the big thing. I have talked so much for a long time, and now they have decided something had to be done about it. The Chief Commissioner and the Director of Veterinary Services conjured up a ridiculous £10,000 livestock improvement scheme, in company with His Excellency the Governor, last March. It was very hush-hush, and the Department of Agriculture was not asked its opinion. A most extraordinary situation has arisen because this scheme cuts right across my training, as regards genetics, (breeding from crossbred bulls) and is also incompatible with Government's 'mixed farming' policy (on which I am supposed to be an expert). It envisages ranching conditions instead of conditions of settled agriculture. I have been carefully schooling all the political people with whom I come into contact, and sought an interview with His Honour the Chief Commissioner on my arrival in Tamale. He has been led up the garden path by the Director of Veterinary Services. A dispatch has come back from the Colonial Office asking why we were not consulted. The Chief Commissioner is committed up to the hilt, and says he personally agrees with me, but the Director of Agriculture won't move. What a state of affairs!

125

With a lot of moral support I surprised myself and the Conference, by countering the paper of the Director of Veterinary Services upon this scheme. So I usurped the powers of the Director of Agriculture; but I was careful enough to preface my remarks with their being my own personally, and not those of my Department necessarily. I had the backing of the whole conference bar the Vets. And CC whom I know agrees with me. Heaven knows where the upshot will be. Once the ice was broken I enjoyed myself, and for a whole morning from 8.30 am till 1 pm we debated. The day before, I was drawn into the conference on the subject of 'Health,' (and not sanitation) and our methods of making farmyard manure in 'mixed farming'. We gained a lot of ground from Dr Lenox, the Senior Health Officer, who read the paper. He is a friend of mine, and we went over the ground the night before. Then I read my paper upon 'Mixed Farming in Northern Nigeria,' with particular reference to the part played by Native Administration. I felt terrible before the Conference, but elated after. I'm afraid I talked a terrible lot. But I knew what I was talking about, and have a fairly detailed knowledge of things up here now. Things must move now I am certain that the prestige of our Department has gone up, and I am glad.

Today, I handed in my bulletin upon 'Mixed Farming' to the clerk for final typing so that is another important milestone and a millstone off my chest.

But still the War goes on in Spain and the last King of England was King George V. Local affairs here have crowded everything out of my mind lately. I have read no papers, and done nothing but write and talk. Everyone here is very displeased with Edward. He gets £150,000 a year to be King, and even I am Victorian enough to deprecate a second divorcee.

I am apprehensive for the future, but refuse to join the Local Defence Force until it is made compulsory. Perhaps you will agree with this? I refuse to be party to fictitious paper strength. I cannot become an efficient soldier in a fortnight's attachment. Neither do I believe that to spend millions upon instruments to kill is compatible with Christian principles.

I am still staying with Gordon since there is no decent bungalow available in Tamale at the moment. It is very pleasant, he is much better in health now than he was. I shall probably return to Zuarungu about the end of the month…"

It was time for me to head north again to Zuarungu at the end of January, my annual report was finished and my bulletin on a visit to Nigeria had been sent home to Waterlows for printing. I had at last achieved an office building in Zuarungu after five years; it was built departmentally of mud bricks with a thatched roof, three rooms and a veranda for the sum of £20. The farm buildings were extended to accommodate an increase in livestock and a seed drying floor. This was of a convenient size for a tennis court, made from anthill earth, dung and 'dorowa' beaten down by local ladies.

11th January 1937 Still in Tamale "Mother dear,

Have I thanked you for the superb Christmas pudding, there was vocal appreciation from us all at Daboya at the New Year. With our little group we took off from Thursday evening to Sunday morning, and had a grand time on the river. Gordon has a little boat which he takes on a trailer behind his car, and a little outboard engine. I preferred a dug out canoe – with three chaps paddling and a camp chair to sit in.

We caught no fish, and I was sorry for Gordon. I enjoyed the harmattan cold and dryness. We wore a sweater right up to lunchtime as it was so cold. This made a fine break and got rid of the last shreds of ill feeling and thoughts which might have been left after the Conference. On Sunday, I was fed up with fishing, and went off with a hunter to explore a lagoon about four miles from the river, with a view to fishing there in the future. It was about three miles long and a mile wide, but I counted no less than seven hippos, which kept bobbing up and snorting and bellowing like mighty bulls. Around the lake there were grand patches of grass, where the hippo had fed and kept down. It was very pleasant, and I enjoyed my walk. It is fine to have so much country all to oneself in these crowded days.

My annual report is half finished and about a week will see it through. Then back to Zuarungu as soon as possible, for things are growing and my office is foundation high with paperwork.

The Crystal Palace fire must have been a blaze. Was it an engineered job do you wonder? I have such a poor opinion of big business these days that I can conceive of nothing too foul for it to do. That private individuals should make profits out of implements of war, or even war itself, is a sickening and miserable sign of the times. I'm afraid conscription will have to be introduced before I shall offer my services, in a War that at least is made possible by international capitalists.

We had His Excellency the Governor, Sir Arnold Hodson, up here last Tuesday. He is a poor thing too. How he became the Governor heavens only knows. He just waffled a lot about uninformed topics, and he was simply not interested. What with His Excellency like this, our own Director an acknowledged washout, we do not stand much chance of getting our cattle ideas through for a long time to come…"

13th January 1937 in Tamale "Dear Mother,

My annual report progresses, it's just 'donkey work' – figures, figures, figures. But with Barlow's Tables, a slide rule and 'the Beetle's' adding machine, I am getting along at something over an experiment a day. Since all are 23, and I have finished nine, I hope to have completed nearly everything by the end of the month.

I want to get back to Zuarungu soon. I have so much building going on there. My office I hear is 5 foot high now. Two new wells in Navrongo are drawing water at 30 foot and so on. My bungalow is also having a new roof – corrugated iron with thatch on top to make it sunproof and cool. It should keep the rain out this coming rainy season.

My increment form arrived yesterday, which is a good thing since it involved getting over a 'bar' for which various things are needed, including the publication of results of original research. It is also the passing of a Native language exam, knowledge of the Agricultural Ordinances and so on. Unfortunately, it also carries me into another scale of payments for the Widows and Orphans Pension Scheme. I now pay no less than £36 a year to it, and only get half back if and when I retire a bachelor. A bit 'ard!

I am still 'doubled up' with good Gordon, since there is still no bungalow available. When I applied for one Rake, the Assistant Chief Commissioner suggested that I should go and stay with him. But, kind as it is, it is no good for me. I like to have my own place and run my own servants to look after me, things done how and when I like them. One gets selfish like this in this country…"

2nd February 1937 Still in Tamale "Dear Mother,

Apparently I was depressed when I wrote to you on 16th December. Well I am feeling in cracking form now, and return to Zuarungu tomorrow. It was just a stage in the tour, complete with this livestock worry which I felt very deeply. I do not give a blast about it now. If Government likes to throw £10,000 down the drain in the name of cattle improvement, it can never be said that I did not even exceed my position in damning it. Actually, things are going very well just now here. Kibi Jones, the Chief Commissioner, has put through a car allowance for me, and given me £40 to buy furniture for the bungalow at Zuarungu. He promised another £20 next financial year. So I have bought a new long bath, another armchair, a settee, three coffee stools, boot rack, towel rail, lamp stands, filter, wardrobe and more dining chairs. So it looks as if I shall have one of the best furnished bungalows in the Northern Province very soon!

Marman started his 100- mile walk up the road this morning; he does it in easy stages over a period of five days. He is in grand form. I am getting a tennis net made at the Prison, and will noise it abroad that anyone coming to Zuarungu must bring a racket, so I may get some games.

I have had a plate bolted under the cracked chassis part of the car, and I am hoping it will last the pace till I go on leave. It will have served me well if it does…"

Work in the district was going ahead with Native Administration farms at Bongo and at Garu; these were family units of 10 acres worked with a pair of oxen. I agreed to write another bulletin summarising four years' work entitled "Agriculture in North Mamprusi" to be finished before I went on leave in May. The campaign to control grass burning was showing signs of success and suggested that Native Administration bye-laws would serve a useful purpose. The advantages were considerably more grazing, less erosion, less damage to roads, better tree growth and improved water supplies. These were results which could be seen almost immediately and could be achieved by the whole population. It was interesting that we should be achieving results after nearly five years' work. This seemed to be the minimum period within which one could see results in agriculture. Once the ball of extension started to roll, and the confidence of the people was obtained, there was no limit to the possibilities provided programmes were comprehensive and soundly based.

12th February 1937 Zuarungu "Mother dear,
Life jogs along here very much the same as ever, and I am pleased to be able to say, that it is placidly on an even level with every step a step forward, in what I feel to be the right direction. I have been very busy since I returned to Zuarungu, for we have a big programme of capital works to get through. This is rather a turning point in our work. We started off my surveying to find out what the people did first. Then experiments and now experiments and extension work.

My office is going up and is wall plate high. It is going to be rather good I think, with a big window looking out to the Tong Hills, a clerk's room and a store and field laboratory. The front is to be built of laterite stone, and the main body of the thing with mud brick walls 2 ft thick, and a corrugated iron and thatched roof on top. Then I am acquiring another 30 acres of land in Zuarungu, and enlarging the scope of the whole place.

I got back from two days trek to Bongo today, north of here, where I am starting a Native Administration demonstration farm. This is a completely new idea in the Gold Coast, though a common practice in Nigeria. I am starting another of these farms at Garu in Kusasi 50 miles away.

I have just had an interesting showdown with the Ministry of Health in Navrongo over our methods of making farmyard manure, and its dangers for fly breeding. We have won the day, and far from condemning our methods, they are going to encourage them and thus our perpetual enemies are 'being turned' into our best friends.

A return of the harmattan has given us temporary reprieve from the stifling heat we may expect in March and early April, and of which we had a taste just before I came up from Tamale. Cold at night, two blankets and a dressing gown, and one is driven in to the warmth of the house to sip tea at 6 am..."

25th February 1937 Zuarungu "Dear Mother,

The business in everything seems to continue unabated, and it looks as if I shall be 'all out' for the rest of this tour. I got back from trek in Kusasi, that is Bawku district due east of here, where I am putting a new demonstration farm this year paid for by the Native Administration. I met Kibi Jones with an entourage of Coastal dignitaries, visiting Bawku. (136) I found that Steel and Mr and Mrs Auchinleck had passed up to Navrongo, and returned to Tamale today.

Whether any good can come of this visit it is not possible for me to say. He talked and listened to more 'agriculture' than ever before, and was more reasonable than usual. He intended to give me a raspberry, so I heard, but has gone away without doing so. I have had an official fight with the Veterinary Department this year – at the Conference upon the cattle breeding policy. Also, with the Sanitary Authorities upon sanitation in Zuarungu, and farmyard manure; and to have to fight one's own Department would be just too much. Auchinleck made Gordon pretty miserable in Tamale by the silly things he said, and I suspected trouble and was prepared.

I have every reason to think that he was not dissatisfied with the work up here. My report upon 'A visit to Northern Nigeria to study mixed farming', has gone home to Waterlows for printing as a Departmental Bulletin, and now I am pledged to writing up the four years' work up here before I go on leave, in the form of another Bulletin. So I reckon I shall have done a reasonable tour's work by the time I step on board the M.V. Apapa.

Bought a grand little cow and heifer calf last week, which I hope is the first unit in what will ultimately be a pure bred herd of local cattle. Zuarungu looks more and more like a proper farm every day.

My campaign against the burning of 'bush' grass has been pretty successful in this area this season. I am sure it will be a great help in checking the awful soil erosion, which normally goes on here. Also it fits in with our

130

policy of cutting grass for hay or bedding for livestock. The trouble is to get our agricultural policy not too many years ahead of the social development of the population. (137) It is a most difficult thing to gauge. Anyway we had the first rain of the year today, and it has been very humid and hot. It is very cloudy now.

But I don't want rain yet. We are just building up the gables of my new office with mud – unburnt bricks – and by the curious workings of Government, we cannot afford to pay for timber for rafters and corrugated iron until next financial year. Since we don't know if the estimates have been approved we cannot order ahead. It is going to be a very substantial building and will be a great asset…"

13th March, 1937 Zuarungu "Dear Mother,

Zuarungu has been a tourist centre for some time now. I have had various people staying with me on and off for the last two weeks. Their entertainment coupled with a busy time in the office, and on the farm, has not left much time for letter writing. I had Gordon Cowan up for the last few days, which was extremely pleasant and we discussed many and weighty things. We are both committed to writing Bulletins, he upon the stuff in the Southern Province of the Northern Territories, and I for the Northern Province stuff. I promised Auchinleck that mine would be ready for the printer before I went on leave. But this is a bad season for concentration for a number of reasons. For one thing, the weather is trying; we are just in the transition stage between the dry season and rains, when the wind does not know whether to blow from the N.E. dry or the S.W. moist. This leads to dust devils which may sweep through the house, leaving it incredibly filthy. The temper wears thin at this season and stage in the tour.

Then, agriculturally, it is a rather exciting time, getting plots marked out for next year's experiments, training men and bullocks and getting everything ready to take immediate advantage of a rain-softened soil. Then my office, made of mud bricks, has no roof as yet and the walls are not proofed. I hate to think of the destruction 2" of rain in a tornado could do to it, in its present condition; and it is a probability. The finish of the annual report, and the usual scramble at the end of the financial year, also seems to keep the mind occupied…"

7th April, 1937 Zuarungu "Dear Mother,

But I get very little time for reading. I am still slogging away on the Bulletin, but something always keeps bobbing up just when I think I can put in three hours (For example, where are we to build the mixed farmer's house? The carpenter wants four nails. The well has gone dry. What work is the mason to do tomorrow? Voucher forms are finished. Samples of manure for the chemist. Early millet seed from Nigeria. And so it goes on.) But it is a great life and this year has seen big changes. This morning I was over in Navrongo, to address the Chiefs of the Kasena-Nankani Federation, and to give a practical demonstration of mixed farming. It went down very well, or seemed to. Another boost for agriculture.

My office is finished but for the doors and windows, and we cannot make these until we get some timber up from the south. It is a nice little office. Now I am waiting to build an implement factory.

I am feeling reasonably fit, but very tired and wanting a holiday. Eighteen months on end competing with a lowering climate and the problems, takes it out of me. I have had a surfeit of 'shop' this year. I cannot keep out of things; hence I get mixed up in sanitation palavers, supply of firewood to the mine, timber concessions, Native Administration Schools, N.'s, water supplies and so on. I find myself in inter- departmental brawls in no time.

Our anti-grass burning campaign has been remarkably successful in this area this year, and I am certain that everyone will soon realise the value of it. It is only when the bulk of the people don't burn, because they don't want to, that we shall get real control. I have foretold more grazing and stock fodder generally, less soil erosion, less damage to roads, better growth and fruiting of trees and a higher average height of the water table…"

In April 1937, Meyer and Sonia Fortes left Tongo after his anthropological study of the Tallensi. (138) We kept in contact for many years, exchanging letters and on visits to England.

17th April 1937 Zuarungu. "Dear Mother,

As I write, the long looked-for rain approaches. It is very late this year which means that my anti-burning campaign has been exercised. I hear a mighty roar in the distance, and the sky full of lightning. But it will clear the air. The weather has been fairly trying these last few days; although I must confess that I have noticed this more from its effects upon others, than by my personal reactions. For I have been feeling very fit, and got my second wind at a time when I most needed it.

I see Auchinleck has put a minute on my application for leave, asking for an assurance that my Bulletin on North Mamprusi is ready for the printer before I go. Well, it progresses if slowly, and I slog away at it literally morning, noon and night. But it is not a thing one can rush.

Auchinleck has written a super memorandum to the Colonial Secretary about our work and future developments from the NT, showing that our arguments have 'gone home'. Pity he didn't give us the satisfaction whilst he was here, of knowing the way he was thinking, and then I at least would have been less hard on him. But then I might not have been quite as frank as I was, and this was part of his game.

Now we are being almost embarrassed by the attention we are getting. But five years of sound work, or work on sound lines, is standing us in good stead, and I at least, have full confidence in the future. Locally that is.

I wish I could feel so sanguine about world affairs. I think they are very unhealthy just now, but I have had so little time of late to think about things – which are just as well -– that I am not as troubled as I was.

Gordon Cowan is coming up on the 23rd to 'take over' my beloved Zuarungu. This will give me time to concentrate on my magnum opus.

In a month I shall be creeping down the road, staying in Tamale and then on to Kumasi for a couple of nights with James Broach. Then I am going to spend a couple of days with John Symond in Accra, and sail from there instead of Takoradi. Symond is acting Agricultural Director just now.

Then on that restless sea and a peaceful pipe in my long chair, now a little read, now a sip of lager, now a long gaze at the sea. And I shall be able to look back upon a tour during which many things have grown up, and much progress has been made. I am still tasting and savouring my Nigerian trip, which was indeed a super show from every point of view. It has been of inestimable value during the last few months. It clinches an argument to say 'well, in Northern Nigeria they do so and so'…"

I was looking forward to Gordon Cowan coming up to relieve me on the 23rd April which would have allowed me to concentrate on my Bulletin which was promised before I went on leave in May. But alas, Gordon was taken instead into Tamale hospital with a suspected duodenal ulcer, and I was called down as his closest friend to try to comfort him.

4th May 1937 Tamale "Mother dear,

Just a hurried note to tell you that I was called to Tamale last weekend to see Cowan, who is lying here very, very sick. Poor old man has scarcely eaten anything for two weeks, and is appallingly weak. I want to stay with him and see him home. I have been busy packing up his house during the last three days.

Tomorrow, or the day after, I return to Zuarungu to pack up my stuff, and then come straight back to stay here and await developments. The point is that I may extend my tour by two weeks, in order to let Gordon get back his strength. Nothing matters but to get him fit anyway.It is as well I am feeling very fit myself.

So mother dear, do not be surprised if I am a fortnight late; indeed, it will mean that at least Gordon has lasted so long. Tamale is singularly lacking in facilities, and is a depressing place in which to be sick .

Now I must go and bathe, and go up to the hospital. Will leave this open, but may have to close it in a hurry.

Wednesday morning.

G.C. was considerably better last evening and had his first square meal at lunch yesterday – a little egg and potato. I believe I have a good deal of influence over the old man – of course we are very close friends and can talk with complete freedom. Sando begged me last night not to go to Zuarungu yet because 'My Massa no do talk, unless you stay and he no do get better'.

Well, now we are into May and it is a lovely month. No rain in Zuarungu when I left but it is lovely and green here now in Tamale and the mornings sparkle.

Alas, my Bulletin which must now take its chance. I have got more than half done and the rest can be done anytime. Well, mi' dears. God bless you. I shall be very, very soon in your midst and if in the meantime I can assist in making Gordon strong again, well it will be just grand…"

11th May 1937 Tamale "Mother dear,

I hope you will not be disappointed but I have had to extend my tour by a few days, and will be sailing, in all probability, by the MV 'Accra' leaving Accra on 13th June and scheduled in Plymouth on 25th June .

It has been a terrible 10 days. Poor old Gordon, we had watched up and down the whole week until Saturday, 8th May , he was operated upon. The stomach had perforated due to the gastritis, and an ulcer had formed. He came round after the operation, and I spoke with him at 1 pm. I told him the doctors were pleased with the operation, and he said 'Clever chaps aren't they' and smiled. At 3.30 Dr Cooper called me, and I stayed with Gordon until he passed quietly over, at about 5.15. It has cast a terrible gloom over Tamale, for Gordon was one of the finest men ever to draw breath. And now he has passed on, very quietly, at the premature age of 33, and I have lost one of the best friends I ever had. (139) He has left a big blank.

I return to Zuarungu tomorrow. I feel like being alone for a bit, to re-arrange my ideas. I have much to do, and shall be busy up till the very moment I go. I am a bit mixed up at the moment. I may soon be in Tamale again, as executor for Gordon's estate. I have already packed up his effects. Now I go and pack mine…"

He battled for 10 days and on 8th May Dr Don came up from Kumasi to support Dr Cooper, and they operated. He had been stationed there for eight years; he was a great sportsman, a fine agriculturist and a very gentle man, greatly loved by so many who had been privileged to know him. There was no X-ray apparatus in Tamale and in those days no air service. A further tragedy was the death of two nursing sisters on the road coming north at night to assist. The car in which they were travelling ran into the back of a lorry, invisible in a cloud of dust, a sad waste one felt should not have happened.

I finished writing my Bulletin "Agriculture in North Mamprusi", which was published as Bulletin No. 34 of the Department In due course, I received commendation from the Colonial Advisory Council in London. In the meantime, an excellent memorandum was written by my Director to the Colonial Secretary in Accra, outlining the policy for the North. He confirmed our desire for continuity; what a pity we did not have this backing five years earlier.

I went on leave later than I had intended and sailed from Accra on the R.M.S. Accra on June 13th 1937, and arrived in Plymouth on June 26th. This was a fateful journey for me because travelling on this ship was Marjorie Beatrice Barnard, who had been the Official Reporter for the Nigerian Legislative Council and Confidential Secretary to the Governor, Bourdillon, in Nigeria. We became engaged and were married on 28th August at St Clement Danes in the Strand. So much for all the brave talk amongst my colleagues about rats leaving sinking ships, but at least I had seven years' service and was confirmed in my appointment.

Marjorie was the youngest of three sisters, who all worked in a secretarial capacity in the City of London. She had worked for the legal firm of Bentley & Co. – W.O. Bentley, of Bentley car fame, would sometimes come into the office to see his elder brother. In 1936, she felt the urge to wander and put her skills to greater advantage. She learnt to drive and took the Civil Service exam before applying for the post in Nigeria.

Sailing from Liverpool on the Abosso, she arrived in Lagos two weeks later. She was given a wooden house, raised up on stilts, overlooking the harbour and near the Secretariat where she worked. Here Marjorie had to adjust to an independent style of living. Her steward boy, Knockoff, cooked, cleaned and shopped for her in the local market. She had to adjust to being on her own in a very different country, which was quite a daunting task.

Her job proved no less difficult as she had to take down dictation from all members in the Legislative Council, including local dignitaries, some of whom were difficult to hear and understand. The work in the Council was equivalent to Hansard in the House of Commons, and all spoken words were taken in shorthand and typed up for publication. Being a good looking single woman of 32 she had ready help from many beaux! One thing that stood out in her mind was a midnight call to the Legislative Council to report an Extraordinary Meeting. The Governor had to announce the Abdication of King Edward V111. Many Chiefs attended in their fine robes, and left tears pouring down their faces saying their King had been "destooled".

One of her beaux sent her a wonderful poem, on hearing of her marriage. "The rumour has started, the news has spread,

That you are changing the place where you lay your head,
Deserting your protector in Old Forces Road,
For a neighbourhood that is more a la mode,
My sword to its scabbard I sadly return,
For no more will I draw it with look grim and stern,
To escort you in safety as the road you cross,
To return to your house and a well earned doss.
Yours ever, Fiery (Faison)

There was an interesting Colonial Summer School held at Oxford in July attended by a strong Gold Coast contingent, George Gibbs, John Armstrong, Harold Blair and myself. We were delighted to have the opportunity to meet such colonial luminaries as Lord Lugard, Professor Coupland and Margery Perham.

Marjorie Lynn 1938   Ploughing 1937

Zuarungu simple Cattle pen for cows, 1937

Chapter 6 - 5th Tour 1937–39

Married life and extension work, Anti- Burning Regulations and Manga Station Development

I sailed back to the Coast with Marjorie, leaving Liverpool on the R.M.S. Apapa on 17th November, and we called at Madeira, which was bright and warm just a few days after leaving a cold Britain. The next day into Las Palmas for fuel, with time to go along to the "Accra", tied up next to us homeward bound to see if there was anyone we knew on board. What a revelation to see so many pale, weary and hot- looking people and to realise that we must have looked like that six months before; certainly West Africa was no health resort for the European. We had pleasant travelling companions; the Symes came as far as Madeira where they were spending the last month of their leave.

During leave I acquired a very comfortable Canadian- built Ford V8 Cabriolet from Wood & Lamberts in Albemarle Street; it had done only 10,000 miles and cost £100! We drove up on the new road to Kumasi, then up the dusty corrugated road to Tamale. Our house had a swarm of bees in the dining room sideboard, a not unusual hazard! Tamale was hot at this time, the harmattan had not started, but the nights cooled down. It was a busy time with experiments to be evaluated and annual reports to write.

From MARJORIE 7th December 1937 Tamale Agricultural Station "Dear Mother,

We had a very good trip out, though we were advised to keep to our cabin for the first two days! We arrived at Takoradi a day early. It was a very hot and dusty journey to Tamale and we were both glad to arrive.

Charles is fit and happy to be back here, since as you know he loves both his work and this place. I think I am going to like it very much – it is like open parkland, though rather dry at present, since we are in the middle of the hot season.

We are at present living in a bungalow in which Charles and Gordon spent a good deal of time. It brings back many painful memories for him, yet it holds very many happy ones. Gordon must have been a grand man, and is very much missed by everyone here. Although I think Charles is very glad to be once again in this house – I think he is comforted by the thought that Gordon is sleeping so near to us.

Later in Zuarungu – I fully agree with Charles that it is grand up here. It is difficult to define exactly where its attraction lies – I think it is in the quiet, peace and spaciousness of the district. Tamale is a very pleasant station, but one hasn't it to oneself, as is the case in Zuarungu.

I expect Charles told you how lovely the Tong Hills look from the front of the house. The bungalow is built on primitive lines, but can be made to look very attractive and I think I shall grow to love it as much as Charles does. We sleep out of doors too, which is very good. There is only the DC here beside ourselves, and even he seems to be on trek for most of the time. So you see it really is remote.

We went round the station yesterday evening. It looked in very good trim, and even Charles is satisfied. It must be fine to see a thing of one's own planning turning out so well. He has some spurwing geese which have grown tremendously.

The natives round here are rather nice to deal with – three of them arrived this morning with a gift of two chickens and a plate of eggs. It is rather a charming custom I think, don't you? One of the old men claims that his father on his deathbed charged him to do no more work, so he daren't do so. (140) I think they get round it by arranging that he shall make manure on his farm for Charles, that isn't working for himself you see. Then I expect you have heard Charles speak of Mr Na – he is a grand old man and he came to greet me in Tamale, one has to shake hands with him. Real local colour!

Quite a number of people have said they were sorry for me here by myself and 'What ever will you do with yourself?' But there are so many things I want to do, that I am sure the days won't be long enough for a considerable time to come; I suppose if you did not like country life one would loathe it here, but to me it is much more satisfying…"

The annual Northern Territories Conference took place in mid-December and was less acrimonious than sometimes. I had two

139

matters I wanted discussed, the first was the acceptance of the desirability for Native Administration Grass-Burning Regulations, and the second the formation of a Northern Territories Board of Agriculture. Both proposals were accepted, and it was left to me to follow them up.

My domestic staff seemed to have accepted their master's married state without demur. However, on Christmas night when we got back from dinner at the Residency, we found the faithful cook Abugari waiting for us with a complaint. Apparently, Anasuri (141) the steward boy, decided to celebrate Christmas with two bottles of my beer and a tin of herrings in tomato sauce. He made a good job of it sitting in the lounge smoking my cigarettes whilst the houseboy played the gramophone! Unfortunately, he was disastrously sick, so I lost a good steward boy.

In January 1938, the annual polo match against Accra was held on the Coast. I was included in the Tamale team and Marjorie had a chance to see more of the country.

From MARJORIE 26th January 1938 Tamale "Dear Mother,

We went down to Accra for the annual Polo Match. The journey down to Accra was a nightmare, as the road was almost like a battlefield. We had to do 5 mph for about 30 miles. We took three days, staying a night at Kete Krachi, and one night at Kpeve with Marden Cook, of whom you doubtlessly have heard. He really is a character – he talks and reads of nothing but horses and dogs, and even has all his cushion covers embroidered with dogs. The amusing part is that nobody has actually seen him on a horse! He keeps a pack of hounds and has had a hunting coat made to wear when he takes them out to the accompaniment of his horn! He tells some simply incredible stories of his experiences in the Sudan, which really give credit to his imagination. Two of them include saving the Sultan's daughter and a night adventure with a leopard.

The whole Colony knows Marden Cook and everyone is very entertained by his stories. His boys have amazing dress – white jodhpurs and turbans; it was a very amusing evening.

My wedding present from Nigeria has arrived and, as Chas says, is definitely one of our assets. It is a very nice silver salver with the signatures of the entire Secretariat engraved thereon. I shall certainly prize it for its associations, I was very grieved to hear though of the death of one of its contributors. He was an absolutely grand man of 33 and it does seem tragic. Charles met him on the Course at Oxford this leave and liked him immensely.

There is a funeral on near here, and they are all playing their flutes on three mournful notes. Sounds like Ali Baba's cave or the Garden of Allah! It makes it a bit difficult to concentrate.

The boys have found a cobra about 4 ft long and a python about 16 ft long just near here. I do hate snakes too, especially when one is sleeping out. We can't keep the dogs at hand in case the hyenas take them. Hyenas (142) give a kind of prolonged bark (a horrid noise) and make a noise like a dog scrunching a bone to attract one's dogs out.

We had a terrible week last week packing up – can you imagine it with an indoor temperature of 95°. We left Tamale on Sunday. When we got about half way, part of our wiring suddenly burnt out. Luckily, a lorry on its way to French Country came along and the driver was able to put the damage right – it was a grand sight. All his passengers descended and watched the proceedings – one or two were very picturesque, in their flowing multi-coloured robes and turbans. They were French Arabs I suppose. We stopped for lunch on the banks of the River Volta, which was quite pleasant.

I expect you have heard on the radio of the cocoa palaver here. There is a complete holdup, and I think the Colony finances are likely to be in a bad state next year. I should think there will be cuts in salary, and allowances most probably. It is a marvellous opportunity for another cocoa- producing country to cover the market. Everyone seems to think it serious…."

Agriculture continued to receive the support of the Administration; the Chief Commissioner brought Canon Grace, now Principal of Achimota, and 22 students round to see the agricultural station. Eustace Rake, Assistant Chief Commissioner Northern Territories and Cecil Amory, District Commissioner Tamale, brought a colourful party of Dagomba chiefs to see oxen ploughing, and cattle kept on straw to make farmyard manure. (143)

Zuarungu was a big test for a wife; there was no running water or electricity, no glazing in the house or locks on the doors. The floors were of beaten mud and cow dung, with the familiar wood burning "Dover" stove in the kitchen. Yet the bush life had a charm of its own. Of course, it was all very well for the man who had his job to absorb his time and interest, but what of the wife? (144) Marjorie developed, amongst other things, a successful poultry unit by the house. But one morning when I returned from my hack around the countryside I sensed something was wrong. My cat Percy had been caught red-handed with a newly hatched chick in his mouth. Marjorie was almost in tears and issued an ultimatum that either

Percy had to go, or she would. This was serious because I had had Percy a long time. I went to my desk on the veranda wondering what to do, when my eye lighted on a 12-bore shotgun cartridge. I had replaced the lead pellets with rock salt to shoot at bats, hanging on the roof of the store downstairs. I thought at least I would give Percy a fright; he meantime had tucked himself in high up on crenellations of the wall between the dining room and bedroom. I took my shotgun from the cupboard and fired up at Percy – the effect was electrical. The house was filled with dust and feathers and the skeletons of innumerable chicks. Marjorie rushed out of the bedroom thinking I had shot myself; the servants came running in thinking I had shot Marjorie. Only Percy was unmoved. Tension had gone with the bang and we roared with hysterical laughter. It was agreed to move the poultry unit to the farm. Marjorie and Percy both stayed, and we all lived happily for a long time thereafter.

On another occasion, we had trouble in the chicken run when a python, gorged with a hen he had just swallowed, could not drag himself either through or over the wire netting. So it just hung there, an easy target for my shot gun.

17th February 1938 Zuarungu. "Mother dear,

I am still mighty busy. I could ill afford my time away in Tamale because Yenli – my right hand man – has been on leave, and his relief is a rather weak little man. Consequently, ordinary seasonal building work was very behindhand when I got back. Extraordinary work, such as the learner mixed farmer's compounds and new wells, were even more behindhand. Then there is cattle to be purchased – a slow business in a country where cattle are wealth and not money. (145)

But it is a great life and I am never as happy as when I am busy from 6 am till bed time at 9.30, there is always something to be seen to. I am becoming immensely conscious of the fact that we are coming to the end of our useful work in this area, until such time as it advances sufficiently to take advantage of what we have to tell them. Who knows whether this may not be until the young boys, going to the Zuarungu Native Administration School (146) just opened, are adults. In the meantime, we must push on to other fields, leaving an African staff here to run demonstration farms and carry out propaganda…"

From MARJORIE 20th February 1938 Zuarungu "Dear Mother,

We awakened a few mornings ago to find that the harmattan had vanished overnight, and the Tong Hills seemed to be right on top of us.

142

Charles has been busy getting things up to date, as Yenli has been away also and all the men get very slack without supervision. However, it is for his own work he is doing this. He has got his annual report finished, which is a weight off his mind.

Last week we went out several times to various villages trying to buy cattle. It really is a very difficult thing to achieve. The Chiefs made all the people bring in their cattle, but when you picked out the ones you wish to buy they just refuse to sell. The standard of values here seems to be:

8 goats = 1 sheep

8 sheep = 1 cow

4 cows = 1 wife.

We have managed to get 2 young bulls and about 2 bullocks and 4 heifer calves.

Charles is getting some boys up from Tamale to start a school farm here. They are to have a piece of land, and do everything including breaking in bullocks. All the implements here are hundreds of years behind the times. For instance, most of the people just have little wooden hoes with which to do all their farming .

A few days ago we went to Bongo, and while Charles was speaking through two interpreters to the Chief, I noticed one small child eating a mouthful of raw grass. All the women brought out their babies to show me. Finally, after about 20 minutes palaver, (147) the Chief presented me with his hat full of eggs, and a fowl. We very seldom have to buy chickens.

It is quite hot even at five o'clock. Last night, we talked with the local retired court interpreter, Bassana, (148) who lives in a compound on the edge of the government area. Though a fair age he is still a tall military figure in his cotton smock and cap. He told me a lot about his life. Apparently, when he was 17 a warring Moshi Chief descended on his village and carried off all the women and children. The men he made join him in slave raiding. He did that for years, until the Government descended on them and took all the young men prisoners. They put them in the army, where in due course he became a sergeant major. He told me of the wars in which he had taken part in 1900–04, ending with the Battle of the Tong Hills in, I think, 1912. He remembered that he led exactly 1,024 men! He was going to England for King George V's Coronation, but was recalled when he got to Sierra Leone (he calls it Seelong!) in order to fight in more wars. He is very old now and has a grand philosophy. He really is a dignified figure in his little short chemise and white cap, and it was rather fascinating to listen to his story in his pidgin English.

They always refer to the calves as "pickins" (babies). The effect of their religious belief in cows means, also, that we never get any beef. Some of the people present a grand sight; their pipes reach half way down their legs and rest on the floor when they are sitting down.

We are starting off from Tongo, at the foot of the Tong Hills, and are walking all the week after that; one of the rest houses is reputed to be haunted, and it is a fact that you often get leopards and hyenas prowling round at night. (149) All I hope is that the lions don't get busy. I shall sleep with my torch in my hand!.."

The months of March and April were hot and trying waiting for the rains to break, so that we could get on with the planting and to bring relief to the livestock. The temperature never went below 84° F in April, and climbed daily to115° F.

From MARJORIE 2nd March 1938 Detokko Rest House (150) "Dear Mother,

And now for our recent activities – we went by car to Tongo at the foot of the Tong Hills, and stayed there two nights. The Chief gave Charles a fowl and me a guinea fowl, as the women are not allowed to eat chicken. (151) Then on Wednesday morning we arose at 4.10 am and started off at 5.15 after a cup of tea, for a seven mile walk across country to Shega. There is no road. The carriers transport any load up to 60 lbs in weight for a penny a mile. We arrived at Shega at 7.30. We are both quick walkers, yet the carriers kept up quite easily. It is very pleasant walking at that time of the day, and you can imagine that breakfast was most welcome.

Shega is Abugari 'country' and his Chief presented Charles with a fowl, me with three fowls and a hat full of eggs (which subsequently proved bad!) He must have thought I was a man-eating tiger I think. The following morning we started off again for a seven mile walk to Zuarungu. We are now, as you can guess, in pretty good walking trim.

Navrongo is our next port of call, after which we went out to Sandema where there is a school farm, which is also very encouraging. It has been going a year and is shaping up excellently. The school is run by the Education Department, and Charles supervises the farm. After a night there, we went on to Chana for two nights. You may have heard Charles mention this rest house before. It is situated right on top of a hill. You have to leave the car at the bottom and then my word what a rocky climb, not very long but very steep. One is amply repaid, though, by the view.

I must say I ,too, like trekking very much, especially the days when we have an early morning tramp. One feels so fit afterwards. In the evenings we walk a good way also, inspecting farms and so on. Then, however, we seem to find it necessary to walk across untold number of ridged fields!

We just heard something rather amusing – a certain clerk asked to be allowed to return home as he had 'raptures in his stomach'.

The people in a nearby compound in Zuarungu are supposed to have been converted – I see they have a little wooden cross stuck on top of their house and also a juju post at the entrance. A chicken seems to have recently been sacrificed on the latter. It seems a strange form of conversion. (152)

Every nook and cranny on the farm now seems to be full of cattle. You would have smiled to have seen us yesterday evening, trying to get a refractory bullock into its shed. It just dug its feet in, and would not budge. Topoga (the farm boy) and Charles were pulling it in by the horns and I was administering shrewd blows on its rear.

Then there has been a grand 'to-do' in Kumasi – the Chiefs have presented the Regiment with new drums, so all the Chiefs from these parts went down to witness the Governor formally presenting them. Many of the Chiefs have never been further south than Tamale..."

8th March 1938 Zuarungu "Mother dear,

We are having pleasant weather for March; it has been really bitterly cold some mornings. The chaps with their cattle don't appear until 9 am instead of at six. It came as a surprise to me therefore to find that the mean maximum temperature recorded at Zuarungu in February was 102°, with 110° recorded on February 23rd and a mean minimum of 71°.

I am trekking as much as possible to buy cattle, both for Zuarungu and for Tamale. It is great fun and I am enjoying it. It is a thing I have wanted to do for years. So far I have got about three cows, 10 nice heifers, three bulls and four bullocks. Now I want about six bullocks, and a dozen cows or heifers for Tamale, and have just been sent £120 to spend this month. It is not so easy to buy stock as you might imagine. An order goes to the Chief to produce everything in the morning, maybe 500 or 1,000 animals. I go round and note any animal I like and treat with the owner. I get about 10% successes. The most fantastic answers are given for not selling. Generally, excuses range from the cow was bought with the produce from the farm, and its sale would result in the sterility of the farm, the owner's wives, and his own death.

At least we have got our Draft Ordinance for the Control of Grass Burning into shape. I feel that if I do nothing more for the rest of my service, this will have justified my salary. But I have difficulty still in driving home the fundamental importance of it all.

My next gag is enclosure. This is the natural outcome of keeping selected stock, fostered also by the reading of Lord Earle's excellent book 'English Farming past and present.' I cannot think how it was that I had not studied the history of our own agricultural development more closely in the past, before coming to study somebody else's.

Another pet I am fashioning is a farm school in Zuarungu. Two 'learners' have arrived ready for a two year course in mixed farming. The idea is to give them a miniature holding to farm for themselves. I will treat, in simple lectures, cash operations as it comes up on the farm. It will take a long time to work out to perfection. Well, everything is exceedingly dry still and, whilst from the point of view of my many roofless buildings, I do not want rain. Yet from the point of view of all things that grow, a rest from the strain of this dryness would be a relief. Incidentally, too, it would be a good thing from the point of view of saving the grass, protected successfully so far this season…"

From MARJORIE 21st March 1938 Zuarungu "Mother dear,

Since I wrote last we have been on trek for eight days and bought 24 cattle, some for here and some for Tamale. It was much easier this time, as in the Kusasi District there is a man in charge of all the cattle, and he went round with us.

We stayed at four places, one of them being Wokambo, formally German Togoland. From here we decided to visit a certain turtle pond, which was supposed to be worth a visit. It meant walking about 14 miles. We started off with a guide at 6 am and arrived at the pond at 8.30. It was a very rough road, mostly through uninhabited country. In spite of this, when we arrived at the pond, hundreds of natives suddenly sprang up from behind every blade of grass. The guide said they had probably never seen a white woman before! However, both Charles and I felt decidedly nervous, especially as we couldn't speak to them – we didn't relish being thrown in as a sacrifice to the 'crocs' as it was obviously a fetish pool. I shudder to think what might have happened there. The pool was surrounded by very sheer cliffs. It was a long way down and looked very deep. It was absolutely alive with fish, great and small. Every now and then a turtle would come to the surface, and there were some enormous crocodiles. It was well worth our long walk. We couldn't trace whether it had an outlet. We heard afterwards that it was a fetish fishing day, (153) and all these people were on their way to the appointed place. On these days every man has to leave his work, and the fish that are caught are sacrificed to the juju.

We had breakfast nearby in a dried- up river bed, and then started out the seven miles walk back, and you can imagine it was hot work then.

Whilst we were on trek we also had two bad storms, but now it seems hotter than ever, which I expect is due to the extra humidity. I gather we have been lucky not having the whole of March like this. In the Kusasi District, we saw the men going off hunting with their bows and arrows. They just wear a skin across their backs and have in addition an iron ring and a bracelet with a little knife on it – they use these if their arrows miss the animal and it charges.

The cocoa strike drags on – I expect you have seen it in the papers that a Commissioner Enquiry is being sent out to investigate – I should not wonder if His Excellency gets a reprimand for not anticipating such things. It was a very unsatisfactory state of affairs before.

An amusing incident occurred the other day while we were on trek. It was after the rain and we came to a low- lying piece of the road which had flooded. We sent the boy across, and he reported that the water was not very deep and the bottom reasonably firm. But we got half way across, and then became submerged into a deep hole with the water over the running boards. We sent the boys off for help, and soon about 200 people from a neighbouring village (154) (where it happened to be market day) appeared on the horizon. It is an amazing sight – they were led by the Chief on horseback. The Chief said that we should get out and his people would carry the car across. We started to climb out and suddenly three hefty men seized hold of me, and carried me over. Charles was so overcome by the utter surprise on my face that all he could do was laugh!

Then all the natives did a lot of screeching in unison and lifted the car out bodily, and carried it over. The trailer followed suit. I may mention that both were well loaded. Finally, they all did a war dance round the car– the din was terrific. Of course, we hadn't the camera available…"

23rd March 1938 Zuarungu "Dear Mother,

But we have been greatly spared this year, for only in the last few days would one say this is like typical March weather. For the rest, although we have recorded mean maximum shade temperatures of around 105° , it has been dry and therefore pleasant.

We have 50 head of cattle on the farm, but half of these are destined for Tamale. This weather is embarrassing our fodder supply and accommodation, but we benefit from the manure.

This afternoon, instead of sleeping, we go with Kerr the DC, up the Tong Hills to witness the final of the Gologo dancing. (155) This is an annual custom which I must have described before. It is a fertility dance to bring rain and is cleverly timed to fall on the waning moon in March. The answer in the form of rain is almost certain within a few days. Judging from the sky last night, they will not be far out this year. Cynically, I suggest that the energy they put into dancing themselves silly could better be spent in farming operations. But recent events in Europe suggest there is nothing rational about the 'white' man.

The Gologo dancing yesterday afternoon was very good, but it was very hot up on the hill. Scarcely had we got back when lightning appeared in the sky and at 10 pm a few drops of rain appeared. Wonderful juju…"

From MARJORIE 29th March 1938 "Dear Mother,

Last week we climbed the Tong Hills with Mr Kerr to see the Gologo Dances, which are supposed to call down rain. It was very hot at three o'clock in the afternoon, but Mr Kerr had promised to supply tea on the top of the hills and that kept us going. Judge our disappointment when the cook found he had brought a grand tea but no kettle! However, water is better than nothing. We climbed a huge boulder in order to see the festivities properly, and the stone nearly burnt the soles of our feet. However, the sight was well worth seeing– one can hardly imagine the din – the shouting stamping and singing, combined with the smell of hot natives! To add to the mêlée they all had huge chains on their ankles. We had one of the chants translated – it meant that Dr Fortes, the anthropologist, had come and learnt all their secrets and now everything was spoilt for them. Another song was about one of the ancient wars. The climbdown seemed much rockier, probably because we were more tired. Strangely enough, the same night we did have a small storm"

31st March 1938 Zuarungu "Dear Mother,

The unfortunate occurrence is an outbreak of cerebro spinal meningitis in Bawku, from which place we returned from trek today, instead of tomorrow. I had gone there to buy sheep and a donkey amongst other things.

We are well up to season with a big programme of consolidation, and are just beginning to see some hopeful results in extension work; small beginnings, but definitely encouraging. Last weekend was spent chasing round with the Chief Commissioner and making speeches. I am getting more used to this business now, and don't get so much of a sinking feeling beforehand, and once started, almost enjoy it. But it is not difficult when one has plenty to talk about.

We are laying down the foundations of a herd of local cattle, and I have been busy buying for Tamale too. This interests me very much. Then the farm school is taking shape, and the two learners are getting themselves down to it. I have laid down a small holding for them, upon which they are doing all the work, including looking after the working animals. Later, I hope to make out a course of lectures and notes upon each operation and crop, as it receives attention. After a couple of years these Standard VII boys should know a good deal about mixed farming. It is very pleasing to see one's ideas gradually taking shape. The coming year promises to be even more interesting than before..."

10th April 1938 Zuarungu "Dear Mother,

Work continues to flourish, only wish I could be given more money and facilities. However, we are not doing too badly with what we have got already. Nearly 500 men in Navrongo have been persuaded to make farmyard manure, according to our methods, and we have prospects of establishing two mixed farmers there with ploughs and bullocks, and one at Bawku. These are the first fruits, and very sweet.

There is to be a first meeting of the NT Board of Agriculture at the end of the month, arising out of a suggestion of mine at the Conference. It has come a little early, because I am certain that the Honorary Director of Agriculture will not play. But the principle is right I feel, and to have got the machinery, even if it doesn't function too well for a time, is something.

The holdup of cocoa in the south, which is becoming serious, does not affect us up here very much yet. But the purchasing power of the public is considerably reduced, and therefore consumption and revenue will be down. It is only a matter of time. The Governor's speech at the opening of Legislative Council showed a very flourishing state in the country until this cocoa holdup.

I must give up the struggle. There are no less than five flies sucking the sweat off my knees, and more buzzing around my head. This is the season for the openings of the middens, and the increased humidity brings them out. Unfortunately, I have run out of flit, and it means sending to Kumasi for more, 350 miles, and the earliest it can reach us is next Saturday fortnight – by which time I may be crazy…"

From MARJORIE 14th April 1938 Zuarungu

"Dear Mother,

There isn't very much of note to tell you in this mail. Last week the school master at the Sandema School Farm died suddenly, so that things are upset. They will find it very difficult to replace him, I fear. He was a grand man and there are very few Africans, as yet, educated sufficiently to be able to teach others and also to do good farming work. On his deathbed he disowned all his wives and turned Christian, which was a triumph for the Mission I suppose.

We have both had a typhoid injection, as immunity only lasts for one tour, and it is as well to be on the safe side. So far it hasn't upset us at all…"

From MARJORIE 30th April 1938 "Dear Mother,

On Tuesday morning Charles had to go in to Navrongo, and I went and spent the morning with Mrs Reid, the Dr's wife. I think I told you she is a nice Scots girl, and I shall miss her when she goes home towards the end of next month. The only other woman up here is Mrs Syme at Bawku, and we don't go there very often, not more than two or three times a year.

On Wednesday evening, we went out to the farm at Bongo. As usual the Chief presented me with a chicken and eight eggs. The chicken ran away in the night, and four of the eggs were bad! Well, as we were looking at the farm it suddenly became very overcast, and I decided to go back to the car in case it rained. Charles stayed to finish what he wanted to see. I got halfway to the car, which wasn't very far away, when suddenly I was enveloped in a tremendous sand storm. The wind was so strong that I could only get along a step at a time, and it was just dense dust all around. One couldn't see a thing, hardly the path. Even in the car the wind was forcing its way in. You would have hardly recognised us, as we were both blonds at the end of it. The storm must have lasted for half an hour. It seemed as if the whole surface of the earth must have blown away. It was followed by heavy rain in Navrongo, but here we got about six spots. That night it was very much cooler, but it has reverted to normal once again.

Mr Kerr has departed on local leave, so that we have the station to ourselves for a while. He is going to do a 12 days sea trip – to the Cameroons. He needs a holiday, as he recently had a bout of fever and jaundice. I hear he got about 20 miles down the road and broke a spring in his car, which was a rather bad start.

I just saw a funny sight – I had to visit the small boy's house, and I found he's got all our old 'Time and Tide' there. I don't quite see how he can be interested in World Affairs, seeing that he can hardly understand a word of English! I imagine 'Time and Tide' would appreciate a snap of that…"

After some months of quiet domestic efficiency we suffered a strange upset in our arrangements, when the "small boy" complained that Akugri, the excellent steward boy who had replaced Anasuri, was putting a spell on him, and he feared for his life. I sent for Akugri and to my surprise he carried in his hand a naked sword. He sat quite quietly and handed the weapon over at my request. Suddenly, he seemed to go quite "dopy". That is the only word for this peculiar condition which comes over the African. For the past two or three weeks he seemed to be wandering, with the result that the small boy (who gets 10s a month for doing odd jobs) had to do all the work of the house. On making enquires we found that the boy had been indulging in juju stuff and had, apparently, a house full of tokens – also the faithful Abugari told me that the steward boy had "a bad sickness on his arms and feet Ma". They'll never tell you these things until the boy has gone away – I suppose they think that they might get a juju put on them! I was rather scared on hearing the latter. Meanwhile, the wall- eyed Abugari and the small boy are our mainstay. I suggested he should go away for a couple of weeks holiday but he was never well enough to come back.

The vegetation around Zuarungu was varied and interesting,

apart from the shea butter tree, the dawa- dawa, combretums and terminalia; there were many baobabs, acacia, figs, tamarinds and silk cotton trees, both bombax and citiba, all were protected for local use. I derived special pleasure from a specimen of adenium, a remarkable succulent shrub carrying beautiful pink flowers on the stem in the dry season.

There were over 500 farmers by this time putting down grass and straw as bedding for their livestock. I made a point of always meeting a farmer in his midden, to emphasise its importance. The turning up of a corner of the bedding to see if there were any fly larvae was important. They could be easily killed by turning the bedding with a fork. This increased the heat of fermentation. Fortunately, farmers like talking with their neighbours and boast a little of their successes. We found that in most communities there were key men who did not always come forward at first. When convinced, they would make much more progress in a year. We found, too, that there was a tendency in Navrongo area for "mixed farming" to be associated with the Church of Rome. Many of our early converts were catechists of the White Fathers. Presumably, they were amenable to cultural change. The number of plough farmers was relatively low, and it was good that the system should grow slowly.

In anticipation of the increase in extension work a farm school was started in Zuarungu, financed by the Native Authority. They sponsored pupils to become Native Administration Mixed Farming Instructors. The basic educational qualification was Standard VI, and the course lasted for two years, and was essentially practical in nature. I approached the White Fathers for a reference in respect of one student, and they wrote to say the man was worthless, in fact no better than a common bustard. I enquired from the upstanding youth what the trouble had been and in his words "he had fled through the night midst thunder and lightning not wanting to marry, just loving". Stupidly, I enrolled him, only to find in a very few weeks that the White Fathers were quite right in their assessment! We trained over 10 pupils and most of them turned into fine young men of great value to us.

We continued to enjoy good support from the Administration, which at about this time subtly changed from being Political to Provincial Administration. This attitude received a considerable fillip by the attendance, as observers, of George Gibbs District

Commissioner Mamprusi and Owen Butler of the Secretariat, at the Second West African Agricultural Conference held in Nigeria in May and June 1939. Indeed, the close integration of the work of all departments, for sound development, was shown at the first meeting of the newly formed Northern Territories Board of Agriculture. The Board sat under the chairmanship of the Chief Commissioner. At this time, I had a telephone line extended from Bolgatanga, which was a great convenience. I was amused to be called by the operator who congratulated me on answering the maiden call.

2nd May 1938 Tamale "Dear Mother,

The Board of Agriculture meeting took place on Friday and was a great success. This was the first meeting ever held, and the board was convened at my suggestion last Conference. The Chief Commissioner was President, the Assistant Chief Commissioner, and the Director of Veterinary Services, the Medical Officer, Forestry Officer, Dr Cooper the Geologist and myself were there, with Akenhead our new recruit as Secretary. Then on Saturday the races, a pleasant little meeting with lots of people taking part in them. Tomorrow, we go north again, spending a night at Pong Tamale…"

We also gave homes to two polo ponies, Shell and Humphrey, which were being retired from the Polo Club in Tamale; they made excellent trekking ponies. Humphrey was to remain with me to the end of my time in the Gold Coast. On one occasion when we were met by the Chief of Zongoiri on his boundary, Shell, (at one time a Chief's pony), reacted to the drums and went up on his hind legs to dance,(156) much to Marjorie's discomfiture.

15th May 1938 Bawku From MARJORIE "Dear Mother,

The Bawku mail leaves today I hear, and what a mail! The messenger has to walk 45 miles to Bolgatanga and meets the mail lorry from Navrongo, which catches next Sunday week's boat at Takoradi.

We have been on trek for over a week now, and seem to have covered half of the Gold Coast. I think I told you we had managed to get two ponies from Tamale Polo Club, for a ridiculous price of £2 each. We are very pleased with them, and they will suit our requirements. They don't object to stony paths, or swerve as many horses do. Of course, the car is useless now; the rains have set in and washed away the roads. But I am very glad of a few days rest, as I was very saddle- sore. It takes a while to get hardened to being in the saddle for hours on end in the heat.

At many places the Chiefs came out to meet us with musical accompaniments, indeed, at the last place the Chief's son and Elders came out about six miles. On Saturday, we got caught in a storm, and had to gallop about two miles or so across fields to the headman's house. The dust was horrible, and Charles lost his hat en route. When we arrived we were shown into the headman's best room, and you would have been amused to have seen us crawling in. You have no idea how difficult it is to crawl through a hole 2 ft x 1½ ft off the ground! However, we sat on a skin on the floor and kept dry. They even offered me refreshments in the form of a calabash of groundnuts.

At the moment it is raining steadily, which will fill the rivers up a little more; on our way here we had to wade across a river and cross another in a canoe, hollowed out of the trunk of a tree.

Charles has rather an acquisitive eye, and has collected various unusual plants on our way. He was told that one was a juju plant but, said the man 'it is quite alright, this juju likes being asked for'!

The Symes are on trek just now, but the Medical Officer here, Dr Fowler, is very pleasant and amusing. We are seeing quite a bit of him, as he seems rather lonely. He has only just arrived up here, and his wife won't be out until November. In the rains nobody visits Bawku except just the odd person like us, and it takes a while to get used to the feeling of being completely cut off. The stores problem is difficult too – you just have to arrange to have sufficient goods sent up before the road goes…"

18th May 1938 Zuarungu "Dear Mother,

Heaven knows what the weather is supposed to be here, for we have scarcely had a drop of rain since the middle of last October. All the planting dates are telescoped together, and we shall be planting early millet when we should be planting groundnuts. It has been a testing year, too, for my anti-burning campaign, which has met with remarkable success.

We returned from trekking in Bawku only yesterday. I have been searching fairly carefully for a site for a farm centre. There are many considerations to be kept in mind, in choosing such a place. I have narrowed it down now fairly considerably, and know within a mile or so where I will put it. But so typical of this department, it has voted £900 for a bungalow for a European but not a penny piece for farm buildings, equipment, staff, and so on. I am greatly opposed to rushing madly into building a house before we have lived in the country at all. I would like to build a decent rest house for a year or so, and then decide where to put up a bungalow. The cart before the horse again I fear.

Bought a couple of donkeys, a dozen sheep and some goats in Bawku. So our farming becomes more and more mixed every day, and incidentally more interesting. Our difficulties are that we have been so occupied during the last two years with 'capital' works, such as building the office, farmer's quarters, buying a foundation herd of stock, building cattle sheds, and so on, that we have not had so much time to devote to agricultural matters as I could wish. I am now beset with a fear that I may be moved away, or an old man before I have got things settled. But it is a great life.

Now I must think about preparing for the day. We are off into Navrongo for tonight. Never stop moving. It's a devil trying to run half a dozen farms scattered over an area nearly 100 miles wide. The fact that the farms are only eight acres in area doesn't make much difference. If there were 80 there would still be a lot of work, but not 10 times more."

19th May 1938 Zuarungu "Mother dear,

This is the most extraordinary season – a note from Yenli to tell me that the heifer has calved a female calf. No trouble at all; this livestock makes our work much more interesting. We have had practically no rain yet and nothing is planted and we reckon to have all the cereals in by the first week in April. It is rather a strain this waiting. Also, there is no grass and we have got 37 – now 38 head of stock on the place, a dozen sheep and five goats, three donkeys, and they are all very lean. Fortunately, it is a characteristic of native stock that they can pick up condition when the grass does come. Improved stock from home cannot do this.

Well, we are all feeling pretty limp after a visit from His Honour the Chief Commissioner. He came to open the new Native Administration School here, at Zuarungu, and devoted a good deal of his address to boosting agriculture. Then followed an official visit to the farm, where the assembled multitudes were greatly attracted by our new 'donkey cultivator unit' working. Kibi Jones was in remarkable form and did most of the talking. I don't quite know why we got this big boost, but it is all very good for prestige. I am highly delighted with our status up here at the moment, although the apathy of my own department to our work and problems in the North is little short of amazing. But a change will come, and come soon I feel…" (157)

In the new estimates a sum of £895 was included for an agricultural officer's bungalow in Bawku, this sum to include the furniture. After a considerable search a site was chosen at Manga, two and a half miles from the town of Bawku along a high ridge. This made it possible to construct a road straight through, without culverts or bridges. The house site was high with fine views, and the land which was unoccupied stretched down to a fadama, (water meadow) to give dry season grazing and possible rice cultivation.

19th May 1938 Zuarungu From MARJORIE "Dear Mother,

It is a little cooler now, in spite of the fact that the rains have not yet arrived. Nobody can remember such a late year, and I fear that the people are faced with the prospect of a bad year. They won't be able to get a successful early crop, even if they can eventually plant it, and will have to rely on their late crop. It is a pity, as everything should be so green now and, as it is, the grass is brown instead. Charles is getting worried about his stock, as there is no grass and very little water for them. Also, they have all been to the immunisation camp at Nangodi and, owing to the lack of grazing and their reaction to the injection, they have lost condition terribly. However, we are still hoping that rain will soon arrive. Of course, Charles has much less grass available on the farm owing to his having bought so many cattle, both for himself and Tamale, and having had to feed so much extra stock. I hear that Tamale has had some rain and is looking very green.

We have now some sheep and goats on the farm, and it is getting quite a well- stocked place. Our boy has just informed me that it is a hunter to the north of us stopping the rain. (158) Apparently, he is holding things up until his hunt finishes. (All the natives believe that certain people hold the rain up).

I think I told you that we returned from Bawku last Thursday. Well, on Friday morning off we went to Navrongo. Charles has several farms to attend to around there. In the afternoon, we had to go out to Sekoti, where the Chief Commissioner was to address the Chief and people. Agriculture composed about 75% of the address. It is a very good thing to get public recognition out here, as it makes such an enormous difference to one's dealings with these people. As the Chief of Chana said to Charles the other day – 'You have put me right with the Tamale Governor so, of course, I have to do what you tell me to do.' That is their attitude.

Yesterday morning at 9 am, the new Zuarungu School was opened. Various people came in for it and it was a colourful assembly. All the surrounding Chiefs arrived with their people – the Chiefs had their usual brilliant sunshades in reds, yellows, blues and whites, and they needed them. They all sat on the ground while the Chief Commissioner made a fairly long speech, to which they responded. Following that we all went round the school. The kitchen is a mud building and the cooking is done in a huge black earthenware pot over a fire on the ground – no stove. The food is mostly cooked millet, which is first of all ground down to look like oatmeal. The actual school is made of stone (locally obtained), but it has a concrete floor and a tin roof. The boys live in the usual type of roundhouse with a thatched roof, but these are considerably larger and cleaner than the usual native house. There are 33 boys and two girls. Yesterday, they were decked out in little chemises and looked like little angels! There is to be a farm attached, and the students will learn farming and the 3 R's – also, possibly, hygiene. The headmaster generally wears a pair of shorts and a singlet, but yesterday he was arrayed in a very natty cream suit.

In the evening, Charles and I went to visit one of his pet farmers – Charles follows what the man says by his signs, and he can't understand a word he says. Abagna, the farmer in question, received us and conducted us round without a stitch of clothing on. When he comes down to the Zuarungu Farm he generally wears his best 'go to meeting' smock. These people are all very natural though, and, being brown, one doesn't notice their lack of clothing much…"

From MARJORIE 29th May–3rd June 1938 Zuarungu "Dear Mother,

A heifer and a bull are ill – the Vet says it is snake bite and has given them an injection. The bull will be alright but the heifer is in a sorry state, poor thing. I do hope she lives, as it is a new purchase. They are bottle feeding her. Against this, a new calf arrived – a black and white one. Very sweet and growing quickly.

So we are able to get a little fresh milk for ourselves – it is nice after having tinned milk for months.

We have had half an inch of rain this last week instead of eight inches. Charles has planted in desperation – a month late even now – in the hope that we may get a good rain shortly. They say that we aren't likely to get the usual big rains now. Tamale has had very much more rain, and is looking green and fresh, but here everything is still brown.

Last Sunday, we had to go to Navrongo, spending the night there, and the following day and night at Chana. Next Tuesday, we are off on a 10 days trek up to Bawku, without the car this time as, owing to rain from French country, the drifts over the rivers have become impassable. If we do move to Bawku, that is the part I shan't like, – being cut off, but on the other hand there is a market, and we shall be able to get fresh meat, onions, Eh! practically every day. However, there is bound to be a serious drop in revenue this year, owing to the cocoa hold up, so I suppose there is a possibility that they may not proceed with the Bawku scheme.

Oh, when we were in Navrongo we saw an acrobatic troupe from Monrovia – three tiny girls of about five or six and two muscular men, plus three drummers. They wore a wide shell girdle and a massive headdress also fashioned from shells and feathers. They had a small stuffed or imitation crocodile, and this juju they touched before each turn. It was a good show, and we enjoyed it all the more through not having seen any entertainment for months.

Coming back from Navrongo we found that the road had given way in one place, owing to some rain there, so we had to alight and build ourselves a deviation through the surrounding bush with logs and stones. We unhitched the trailer and tore over the place; luckily we got across without getting stuck. Mr Kerr, however, arrived at the same place shortly afterwards and was misguided enough not to reconnoitre first. He got beautifully bogged down and took ages to get it out. He said afterwards how annoyed he was and yet, as he said, only two years ago he had to take off his clothes and swim across at this place.

Since I started this, we have had some good rain. The other evening we went out to dinner, and just getting back from the next house a terrific storm descended on us. It blows for a good while – full of sand – before the rains come. I was simply terrified, as I really didn't think I could make our house. We only had an ordinary lamp which promptly blew out and, what with the sand, darkness and awful wind I should never have got there without Charles. We then had 1 ½ inches of rain and have since had some more, so some green grass should appear now.

Last weekend, Charles was looking at the Sandema School Farm. I got a bit tired of ploughing over furrows, and thought I would look at the school. I was very amused at the pictures they had up on the walls. There was a very large picture showing all the competitors for the title of 'Beauty Queen of Europe'. Also, a separate picture of the King entitled 'Nab George' (Nab means Chief)…"

It was interesting drawing up a plan for the bungalow at Manga, which was to be built of laterite stone with a light corrugated iron roof, overlaid with thatch for coolness. I received a lot of help from George Stewart, the Provincial Engineer in the Public Works Department in Tamale, especially with the contract specifications. The contract was given to one Busurah from the south, who supplied artisans and labour, and I supplied materials. The intention was to build a Native Administration farm centre on 300 acres of land to serve the Kusasi District, and it was hoped that a newly appointed Agricultural Officer would be sent to the district to take charge. The survey work had already been carried out at Zebilla and the demonstration farms, and a pair of oxen units were in operation at Bawku attached to the school, and at Garu, 12 miles south of Bawku. Work on the building started in the dry season of 1938–39; unfortunately an outbreak of cerebro spinal meningitis placed restrictions on our movements. So from time to time Marjorie and I made our headquarters at the rest house at Bazua, eight miles out of Bawku on the White Volta River, crossing and supervised the building at Manga from there.

15th June 1938 Bawku "Dear Mother,

We have been trekking on ponies every day for a week, and this makes us very fit. It is gorgeous trekking weather.

Last night, I rode out to see our first and only mixed farmer in this area, and was delighted with what I saw. It is remarkable how these folk, or rather this man, has been able to adapt his system to accommodate my new methods. What he can do others will follow. The future for extension work here is rosy. For the man who cannot afford bullocks and plough there will be the donkey and cultivator.

I have been hunting for the ideal farm centre. We have been voted £800 to build a house, but I have got to see to it in this dry season. The stupid Government has voted money for a house and not a half penny for farm buildings, neither have they stated what the chap is going to do who lived in the house. However, I am going along my own lines, those indicated by Stockdale and similar to shows I saw in Nigeria. It is great fun looking for a place to make a house and farm.

I go this morning to inspect soil pits on site No. 2, which is two and a half miles from Bawku; magnificent view in every direction, really superb. Then the land slopes down for 200 acres to swampy land, which will provide us with dry season grazing.

We leave here I think on Friday, for Sapeliga. A place away in the hills I have not been to before; thence to Teshi and Zebilla, Tili and Sekoti where I left the car. Then it will be exciting to get back to Zuarungu and see the dogs again. The crops will be well up and experiments will show change. The seeds in the garden, the new hedges round the paddocks, the weight now of Dolly's calf, have they found water for the school farm well? Life is full to the brim with interesting things…"

30th June 1938 Zuarungu "Dear Mother,

At last the rain, but Zuarungu is still very backward, and many of our pet schemes must wait for another year. The living conditions are extremely pleasant just now. Everywhere gorgeous greens and it is just as difficult to imagine the change from the barrenness of the dry season, as it is to remember what cold feels like in a temperature of 105 °F.

It is not merely a single new season which gives me pleasure either, but rather the sixth new season since I first started here in 1932, and this is long enough for trees and hedges to have grown up, and pastures to become established.

The new grass has improved the stock. Dolly's calf is now a month old, and has more than doubled its original weight, of 25¾ lbs, during this time. Dolly has given us two pints of milk a day as well, which pays a third of the daily wage of the 'mixed farmer'. Two lambs have been born this week. So we progress. Then there is the excitement of designing the house for the new farm centre in Kusasi, which I shall probably have to build in the dry weather, the drawing up of a program, and so on. Talking with Captain Mothersill, the DC in Navrongo this morning, he heartened me very much with great support for developments in the Navrongo Area. He promises support from Native Administration funds for any schemes I like to put up.

Then the cultivator I brought back from Nigeria, or rather the details of the pattern of it, is meeting with great success. I am now starting up mass production in a new implement factory, and making sets of donkey harness – for donkeys can pull these small implements and many people have donkeys, many more than have bullocks.

The extension side of the program is the sweetest...."

North Mamprusi seemed a particularly serene part of the world at the end of 1938, which contrasted sadly with the rapidly deteriorating state of affairs in Europe. About 50 oxen with their owners were in for training in ploughing. The new Native Administration Anti-Burning Regulations were being implemented very successfully in Zuarungu District, where it mattered most. It was agreed that a controlled early burn, along a line demarcated by a bush path, would be made around the whole district. The buildup was going ahead in all areas, to accommodate extra staff to deal with the increased tempo of agricultural activity.

The seed drying floor at Zuarungu was enlarged and levelled to make an excellent tennis court. It was made ostensibly for the pupils at the farm school, who were receiving intensive training in mixed farming. But it proved very popular with all the African staff, my clerk George Ankrah, Nadawlina Yenli, Dorkachoe Addo (the nurse dispenser) as well as the farm pupils and visiting Europeans. It became a focal point and safety valve by providing fine recreation. Much of the progress in North Mamprusi was due to the effective personality of Nadawlina Yenli. He inspired an esprit de corps and enthusiasm in the staff, and confidence in the Chiefs and their people.

22nd August 1938 Zuarungu "Dear Mother,
I'd have written more, but a big noise announced the death of two small

159

snakes and caused much but-but in the compound. Faulkner, the Vet here, has just left after a sundowner to walk over to the rest house. I had to persuade him to wait for the small boy with a hurricane lamp, (Faulkner is in his first term) and I threaten him with snakes. But we don't see many.

I am busy as usual. If it is not the Chief Commissioner it is somebody else. This time it is the Director of Agriculture. We may have to go to Tamale, and the Chief Commissioner has asked us to stay with him at the Ridgeway. My clerk is on leave and my great Yenli is sick, and I have to go and see him into Navrongo hospital tomorrow…"

From MARJORIE 25th August 1938 "Dear Mother,

I expect Charles has told you that we have been on the move for 10 days out of the past fortnight – we went up to Bawku, not an easy thing now, as the two rivers are overfull for our horses to cross safely. We had to cross in canoes and then cycle between the rivers. It was ages since I had been on a bicycle, and a man's machine at that. It wasn't as easy as it would be in England, as they don't bother to keep these 20 miles of road in repair in the rains. One had all the time to be prepared to dismount suddenly. Also, was it hot! Charles was singled out for misfortune as first of all he had to wade into the water to get the puppy out.

Then the carriers succeeded in dropping his clothes and book boxes in the water! So that evening he had to wear borrowed plumes! After that his bicycle was broken twice by the boys. I don't think he deserved all that…"

From MARJORIE 25th September 1938 Zuarungu "Dear Mother,

Since starting the above an awful thing has happened, which quite swamps the importance of the European crisis. Anne, the most loving, faithful and obedient dog in the world, has we fear developed rabies. (159) It seems incredible that such a terrible thing should come upon her, and force such a gentle creature into attacking our other dogs Pete and Ullage, with whom she played for so long. I couldn't bear the thought of Charles having to shoot her, so we have taken her into the Vet at Navrongo, and if he decides after a few days that there is no doubt he will just put her to sleep. She is neither eating nor drinking. We are going to see her today. We had all the other dogs injected some time ago, so we hope that their bites will not affect them. Anne was so thin after her pups that we decided to wait in her case a little while. Of course, she was the one to get bitten. I expect you can imagine how terribly upset we are. Just now I can't think beyond this to what will happen if war breaks out…"

One of the problems of going on leave was to find homes for pets. Maurice Wentworth of the Education Department in Tamale solved the problem for his two dogs, Wilberforce and Ullage.

He told everyone in advance that it was so nice the Lynn's were going to look after them in Zuarungu. So in due course, we asked Maurice when we were going to have the pleasure! Although it was rewarding to keep dogs in West Africa, rabies was enzootic and caused problems. Anne was bitten by a stray, and died from the disease. Marjorie, the Vet Derek Faulkner and I, started a course of 18 abdominal injections. We found, after 10 days, that the serum was some months out of date. We had to start all over again with new serum from Yabba, in Nigeria. Fortunately, we all survived, but it was worrying and uncomfortable at the time.

23rd November 1938 Zuarungu "Mother dear,

A start has been made on the Manga bungalow, the contractor has cut some very nice stones and the soil makes superb bricks. When I was up earlier in the month I marked the foundations, they should soon be finished and then the place should go up like wildfire. At Navrongo, we are building a Native Administration cattle shed, to house animals in for training for mixed farming. We have eight animals there belonging to Chiefs being trained. At Zuarungu we have got 10 animals, and a prospect of 10 more before the dry weather is over. Small but important beginnings.

At the Zuarungu N.A. school, we are making further stock buildings and a compound for the farmers. Alas, our well- digging there has so far been fruitless. Nearer home, we are building a house for the clerk, quarters for more pupils, donkey sheds, bull pens, stables, and a cart shed. Early tomorrow morning I go to the Mission of Roman Catholic White Fathers, to demarcate a farm for them as they too are taking up 'mixed farming'. In the morning I give a lecture to the pupils. You may gather, therefore, that the days are full and interesting, and without good staff we could do very little.

The Anti- Grass Burning Regulations are being well enforced, and the people seem to understand them. I am very pleased with myself for swapping two little heifers for one of the best local bulls I have ever seen. He is a grand beast and a credit to the farm. Where he has been hidden all the time I cannot think…"

In mid- December we went down to Tamale and stayed in the agricultural rest house. The Governor, Sir Arnold Hodson, and the Secretary for Native Affairs, Hugh Thomas, were up for a Durbar of Chiefs from all over the Northern Territories. Teams of school boys from various Native Authorities produced short plays; the Gold Coast people loved plays.

From MARJORIE 5th January 1939 on trek Belungu "Dear Mother,

As you can see we are on the move once more. The last two days we have spent at Tongo, and then this morning we came here on our horses. Tomorrow we go to Winkogo. (160) It is the season of grass burning; this includes trees and anything which is unfortunate enough to be in the way. The African seems to love its destruction. However, this year the burning is supposed to be controlled, that is to say they must burn the real 'bush' and Charles is rushing around to check up. Fire is difficult to control when there is a bit of wind, and then there have been a few disappointments. However, seeing that in previous years they have burnt every blade of grass, anything protected is a victory.

This is a tiny rest house, and I do wish you could see Charles emerging. There is just about enough room to come out on all fours, and one nearly always succeeds in grazing one's arms or getting up too soon and giving one's back a crack. The Chief and his people turned out with their drums to receive us - what a noise!"

On the whole, the Anti- Grass Burning Regulations were being well applied, and already favourable claims were being made of benefit to water supplies. A problem arose regarding the "tengani" or juju groves (161) which harboured the spirits of the ancestors, and particular artefacts, stones, trees, and other material objects can, like animals, become the living presence of the ancestors. If a tree becomes nominated as an ancestor's shrine, his descendants may not eat its fruit. In this way, things in the natural environment become linked to a person's or lineage's moral and ritual observance. They become part of the cult of the ancestors and have semi totemic observance. It was thought they liked to be burnt every year, and this was overcome by the issue of Native Administration permits to the Tendana. He was the custodian, for controlled burns. In fact, it was found in subsequent years that the spirits preferred not to be burnt. Thus, one saw that native custom was not inflexible, but could be adapted to circumstances given time and understanding. The ebony tree was held to be dangerous, and had the ability to magically injure or even kill people.

It was decided Marjorie should go home early, and she sailed in March on the "Abosso". I took some local leave and had a chance to make contact with Headquarters.

At the end of the financial year I purchased from the Public Works Department stores in Tamale, against my vote for the Upkeep and

Equipment of Experiment Stations, a supply of cement with which to replace the mud and dung floor in the Zuarungu bungalow. This was a great success, so much cleaner and less abrasive. When I reported this to Marjorie in England she wrote to say that the Zuarungu bungalow with a cement floor would be heavenly. I didn't know she had felt so deeply about it.

5th April 1939 Zuarungu "Mother dear,

At the moment I am in the throws of rushing around the countryside, to see how the chaps are progressing on the schools, demonstration farms, and our eight mixed farmers. These eight were with me last year, and are working with their own ploughs and bullocks for the first time this year. It represents a terrific change as you can imagine, and they need fostering. Once we have got a number of farmers started it will be alright, for there will be much pride and self help. We had a full 1 ½ inches of rain at the end of March hence all the activity.

The Native Administration estimates have just been finished. I hate estimates. Anyway, every list of money I asked for from Government and the Native Authority has been voted. It is as a result of the visit of the Colonial Secretary last December. So instead of having about £600 to spend (out of an estimate of £900) as last year, they have given me no less than £1,320 of Government money and £1,196 of Native Authority money. It is exclusive of salaries on the General Warrant. It is a fine gesture to Agriculture, but when I prepared the estimates I proposed a staff of two Europeans up here. We shall not get them till I return from leave, if then.

However, much of the money is capital and non recurrent. They are letting me develop Manga as the headquarter station for Kusasi. What an opportunity. I have money for an office, clerk's quarters, overseer houses, stockman's house, cattle sheds, and equipment and so on – everything I asked for. So I'll say we shall be busy putting in foundations and cutting stone between now and leave coming.

Manga house is at last completed and quite fulfils my expectations – which were high. It is not quite Zuarungu however, all terribly new. I had trouble with the contractor, a worthless fellow, as I expected. He drew only £74 (the balance of his contract when the house was finished) which was not enough to pay all his people. The blighter tried to welch, but the carpenters got him to the Chief's court, and he paid them up. The masons he had sent away, on more contract work – septic tanks for the Ministry of Health in Bolgatanga – so they were unlucky, also his labourers. The latter I have undertaken to pay, by purchasing all the materials, which were over-estimated on the bungalow. All very irregular, but what can you do? ..."

21st April 1939 "Mother dear,

I am working at top pressure just now, but being very, very tired I seem to go round in circles. All this expansion is all very well at the beginning of the tour, but later on when one is tired and stale it is not quite so easy to be enthusiastic. We had a little rain yesterday which cleaned the air very much, and we can breathe again. I get very ill tempered and irritable during this season. All wrong for a farmer to become impatient, but I am more of a clerk, surveyor, building contractor, school master than a farmer. I am tired of wandering around. It's just hopeless. I cannot concentrate on the garden, on the chickens or anything nice. I am away for nearly half the month on trek. From the Government's point of view I suppose, they get better value for money in this way. But ridiculously I am alone responsible for the scope of my work. I have had one visit from the Officer in Chief, Tamale, for one night, this year and none from the Director. A subtle compliment perhaps, since I have been given my estimates in full. What a miserable dammed winger I am, as if you have not got enough of your own affairs to think about…"

The new bungalow at Manga was finished in April and we were lucky to find water in the well near the house at a depth of 12 ft. The excellent carpenter cum cabinet maker, Kofi Mensah, from Ashanti stayed on to make the furniture. The timber, odum, like mahogany (Chlorolpha excelsia) also came up from the forest belt.

2nd May 1939 Manga "Dear Mother,

The storm is just banging and blowing itself out. It was very noisy and rough while it lasted, and drove my guests at the drinks party away. My boys are shivering somewhere out the back, and I should think tonight's food is full of sand and smoke. So until some life appears at the back, I'll sit quietly and write to you.

Manga after the storm – Cattle boys are playing tunes on their little wooden flutes; the thatched roof is drip- dripping. Cattle are lowing and dogs barking in the distance. The lightning is almost continuous, but the thunder is further away. My roof is still in place, and I am interested to note Abugari has just produced a saucer of roast groundnuts. (To make up for a bottle of whiskey which broke in the car coming up – due to careless packing I suspect.) The political systems of Europe have much to learn from Nature in the Tropics in this respect – i.e. the real peace after dramatic upheaval…"

14th May 1939 Zuarungu "Dear Mother,

Ever since 2 am this morning we have been having a steady downpour of rain, and how welcome it is. It has come after a fall of 2.7 inches in Zuarungu. Now the seasonal work is well in hand and as regards the farm, we can sit back and breathe again. I really do wish I could say as much for the office side of the work. I am up to my eyes in it. It is a bad year to be away on leave really, with so many capital works to be seen to. I am endeavouring to leave everything as cut and dried as possible for my relief.

Until the 10th of this month I was up at the new bungalow at Manga, near Bawku. I have an excellent little carpenter making furniture for me. It is rather fun designing one's own. I am making it very plain. The layout of the farm there is proceeding, and already things seem to take shape. At Zuarungu, we are still enlarging, and at Navrongo, where I am trekking tomorrow, we are getting a move on with the new farm centre.

All this makes the time slip away, and I am more than ready for my leave.

There is an ugly suggestion that Akenhead may be detained in Tamale on his way back from leave. There is a case against a clerk who pinched some of his money. I am dashed if I am going to stay after 25th June and will get a medical certificate if necessary. The Officer in Charge of Tamale knew last August that he had to make arrangements ahead, and he has made a mess of it. I do not propose to suffer for his foolishness. Anyway, if Akenhead comes straight up after landing on the 13th I cannot possibly hand over to him properly in so short a time…"

2nd June 1939 Zuarungu "Dear Mother,

The routine of the last two days has been suddenly shattered. On the 31st May, I trekked back from Navrongo and had lunch with Parkinson the DC. He was due for a quiet dinner with me that night, to celebrate the Battle of Jutland. Well, the night was a huge success, but not very quiet. Jock Reid – the manager of the Nangodi Gold Mine and Captain Adams and his wife – he is the Senior Pilot at Takoradi– had descended upon Zuarungu for a quick drink and away. Adams is a grand white- haired cheery red faced Sea Captain, as trim as a boat, and very swarthy and all one expects a Captain to be. His wife is his female counterpart, and Jock Reid is a grand character. Well, Abugari came up to scratch; he did not bat an eyelid when I suggested chop for five masters, instead of two. He serenely said 'there is enough' (I have never known it to be different) and wc had a simply grand evening.

Jock Reid has been up and down the Coast since before the War, and has a photographic memory. It was an evening like going through several volumes of Blackwood's Magazines. He told stories of convoys, old Bill so and so, Lighthouses, loads of mules from British Africa to Calcutta, boats going ashore, coming out of Colombo. These sea people are very real, and so sparkling with a twinkling eye. The party broke up at 3 am after some singing.

But Zuarungu has gone mad. The weekends are one thing, but midweek parties are frowned upon. However, Jock Reid's Oldsmobile came down last night to collect Parkinson and I for chop at Nangodi – he had some fresh sausages up from Kumasi.

But this fine cool weather bucks one up. Tomorrow I trek to Bawku. How far up the road I can get is hard to foretell. The roads at this time are very treacherous…."

In June, I handed over control of the Station to Michael Akenhead and went south to sail on the R.M.S. Accra on 26th June . On my way I stayed at the fine rest house at Ejura on top of the escarpment, before leaving the savannah country to enter the forest country. That evening whilst bathing in a canvas tub on the concrete floor I heard a great rumbling which shook the pan roof, attributable I thought to a heavy lorry going down the hill. When I got to Kumasi next day I learnt that there had been a severe earthquake the night before, centred on Accra. It was said that many files in the Secretariat disappeared down cracks in that building, and many strange tales were told of people in various stages of dressing for dinner found wandering about in their gardens. When I asked my cook Abugari what he knew about it he told me that the rest house roof had shaken like a jelly, but he had said nothing to me!

The "Accra" called at Las Palmas to take on fuel, and the German pocket battleship "Gneisnau" was three berths along with naval officers looking through powerful glasses down the quay. I remember registering an icy feeling down my spine as we left harbour and the two ships dipped their flags. War was declared on 3rd September.

Chapter 7 - 6th Tour 1939–41

Wartime, mixed farming, and Lawra, leave in South Africa

I received a call from the local forces in the Gold Coast, and sailed from Southampton on 25th September 1939 on the "Athlone Castle" converted as a troopship. We were in a small fast convoy with the "Franconia" and the "Empress of Australia" with the cruiser "Dragon" as our escort.

Our first port of call was Gibraltar, then with the "Dragon" to Freetown, where we were amazed to see over 100 ships at anchor, including the "Ark Royal." The Germans claimed to have sunk it some days earlier, in the North Sea. I was unexpected, but the authorities in Accra lost little time in sending me back to the North. I had to admit I was likely to be of more use as an agriculturist than as a soldier, particularly at the time of the so-called "phoney war". In Kumasi, I was intrigued to see a bunch of recruits on the parade ground learning English and repeating commands after their Scottish Sergeant Instructor. (162) ("This is ma heed" and "these are ma harnds").

I took the corrugated and dusty road to Tamale and Zuarungu, seemingly getting further and further away from the war. The season in the North had been good, and our programme was pursued with confidence by the staff. It was highly thought of by the people, judging from the numbers who wished to adopt our methods.

I had been back in Zuarungu about a month when I received a telegram from Accra to say there was no objection to my wife joining me, (163) so in mid-December I made the journey down to Kumasi by road, 345 miles.

From MARJORIE 12th February 1940 Bawku "Dear Mother

Please excuse typing as I have forgotten my pen. Really, we move around so much these days that I hardly know whether I am on my head or my heels, and the days simply fly past. We spent two or three days here, then rush on to Zuarungu for a night, and so on. We shall both be glad when we can settle down a bit, but if things go as planned, I can see we shall have to move about a lot in the next few months.

Last week amongst other places, we sojourned for two days in Tongo, a pleasant place right off the main road at the foot of the Tong Hills. This was followed by a night in Zuarungu. We returned here on Friday night, on this occasion bringing the faithful Percy with us. He strongly objected to being tied up in a pillow case for the journey – indeed we only succeeded in getting him in after a struggle. But he now seems to have settled down and to have adopted this house as his home. His only complaint is that, by standing on a chair, we can reach him from off the top of the walls. In Zuarungu he is miles out of our reach and can sit and laugh at us.

One morning last week, we had some terrific pre-breakfast exercise, in the way of fire fighting. Charles has, as I think we told you, got a law passed prohibiting grass burning except in certain 'bush' grass at certain times of the year, but the small boys take keen delight in a good blaze, and are not to be done out of their fun by any law. The people themselves are nearly as childish – if they see a fire on their neighbour's land they must have one of their own. So we had to go out at 6.30 am and organise bands of people to put all the fires out. It is very disappointing for Chas in this area this year, for a large proportion of the grass has gone up in flames. Stern measures are being taken, but that doesn't restore the grass, and it means that one is set back another year. It isn't only the grass; it is all the small trees as well.

According to plan, Mr Roland Smith (164) is coming up to Zuarungu in March, and he shall then probably spend most of his time thereafter here in Bawku, though, of course he will be working under Charles's supervision. Then in July we may be going down to Tamale for a few months, to take over from Mr Broach, who is due for leave. I shouldn't mind going down there for a few months in the rains, I think...."

The continual change round in personnel was a feature of official life in West Africa, necessitated by the incidence of leave every 18 months. A conscious effort was required to ensure continuity of policy, with strict attention to programmes of work. The year 1940 was a year of consolidation so far as mixed farming was concerned. Much time had to be spent on buildings at the three farm centres, Zuarungu, Tono to serve Navrongo district and Manga for Kusasi.

(165) Many farmers were now making farmyard manure, and an increasing number were training their oxen under our guidance. They acquired ploughs and cultivators, with Native Administration loans. The Anti-Grass Burning Regulations were being sensibly applied, and many more wells were being dug and stoned. Continual trekking for some 20 days a month, generally on horseback with carriers, was necessary to ensure progress over such a large area. The area was 100 miles wide, from west of Navrongo to east of Bawku, and about 30 miles deep. Personal contact was important at this stage. I was also aware that it was good for morale to see a familiar European talking about farming, for many of our fighting troops came from the area. (166)

3rd March 1940 Manga "Dear Mother,

I live in such a spate of work these days that all accounts of time are lost, and I'm dashed if I can remember when I last wrote.

Well, during this dry weather I have built immeasurable buildings, and dug many wells, trained farmer's bullocks, rushed around 3000 square miles of country seeing that the grass burning rules are enforced. For some time now we have not lived for longer than seven days in one place. This upsets all routine, both household and otherwise, and our servants and ourselves will be glad of the respite which will come with the end of the financial year.

Of course, I revel in being immersed in work of my own choosing, and only too willingly do I engage upon fresh ventures. It is a fine antidote to the affairs of the troubled world. If I get well ahead with the executive side of things early in my tour, I can sit back later, when I am tired, and rest behind that big word 'consolidation'.

Agriculture in North Mamprusi really is being put upon a reasonably sound basis. We have got a policy based upon local experience, and the sympathy of the powers above. The new bungalow here, and the layout, is most attractive and the farm is developing excellently. A little round guesthouse 20 ft in diameter, with offices attached, is being built now on the other side of our big fir tree. Then the carpenter is making beautiful furniture to our own ideas, and there is no bungalow in the country which is the same in any respect. The chairs are comfortable and big, the tables the height we want them, and so on. However, I'm afraid we cannot develop Manga fully as the headquarters yet. But then we are very fond of Zuarungu, and have great fun planning and laying out this place. I am getting up some assistance, one Smith who is newly out on his first tour and a very quiet lad... I want him to shadow me about for a bit, and then to make his own HQ in Manga.

We had a simply superb day on Thursday at the Gologo Dances. (167) Raced back from Navrongo after paying a visit out there, to lunch at Tongo – six miles south of Zuarungu at the foot of the Tong Hills. Mr Rake was up from Tamale and Mr Guthrie Hall (our new District Commissioner) put up the lunch – he came over from Gambaga. What a lunch too! A beautiful ham, such as I have not seen or tasted since I left the boat, and champagne for those who wanted it. (G.H.will celebrate his birthday on 29th February in 1940, the invitation ran).

It is very hot in February–March and the 29th was no exception; 3 pm was the time for sweaty slumber not hill climbing. But up we went by car, a mile to the foot of the hills, and then a solid climb up the boulder- strewn road to the top. All around hills and boulders looking like a really wild and colossal Valley of Rocks – great lumps of granite, all shapes and sizes. But it was worth it.

I should think a conservative estimate would put the dancers at 5000, and they had been at it for 10 days previously. (They started on the 15th day, after the new moon is first seen.) Armed, cap a pic, with bows, quivers full of arrows, and swords of iron and wood, towels round their heads and loincloths, but nothing on the body otherwise. Each section dances together, six sections in all, wheeling and turning, and always double shuffling in perfect rhythm, to their topical chant. No sergeant majors organise this tattoo, but everything runs smoothly, gradually they string out and run across the arena and reform closer together for the climax. A mock battle between a sort of totem pole, with hawk wings on the top carried by Wachiga section, and Windusi's totem with horse hair on top. And so the Gologo dances finish, leaving one barely enough time to get back to the cars before it's too dark to see the way. We provided the tea. Not for a very long time have I enjoyed a day so comfortably. This is the fifth time I have seen these dances, and I never tire of them. Every year on the same moon, on exactly the same spot, these people dance for rain; wonder of wonders as we came down the hills the thunder rolled and lightning played, and spots of rain were falling as we picked up the lunch loads at Tongo Rest House.

The faithful Abugari made our tea and being a Tallensi, he changed his white drill and brass buttons at Tongo for a goat skin; thus adding true local colour to a gorgeous setting, under a dawa-dawa tree, at the side of the arena. He looks better in a skin than oddments of European raiment. They all do.

This is the hot season and everything pants for rain. Clouds are appearing from the south west and they all roll up the rivers. It may not be for a couple of weeks or more before they come back with a mighty rush and crackle, as tornadoes from the northeast. A big rush now to finish all roofing, and prepare land for planting...."

Roland Smith was appointed and posted to the North, and after an initiation period he went to Manga to take over the programme in Kusasi. In the height of the rains I was surprised one morning when my clerk, Sam Mosu, came into my office. He complained that he was pouring with sweat down one half of his body, whilst the other half was hot and dry. It happened that the Medical Officer from Navrongo "Bats" Sladen was over to pay his weekly visit, and I arranged for him to examine Sam. He was suffering from unilateral sweating, apparently a sign of a highly nervous condition. An early visit to his home on the coast was commended, and a check-up at Korle Bu Hospital near Accra. Sam Mosu had had an unfortunate experience when a sudden tornado one night blew some of the corrugated iron sheets from the roof of his house. What worried him most was the discovery of a bird's nest in the roof with his initials printed on the eggs. A month later Sam Mosu returned from leave, well again.

13th March 1940 Zuarungu "Dear Folks,

Here's a go! I have just defeated a week's indisposition – distemper or flu or something similar – and feel better for the rest. At least we have finished with buildings, of which I have had a surfeit since I returned. Only just in time too, for the rains are upon us.

Now we are enjoying sparkling weather. All traces of the harmattan have been washed away, and we are busy preparing for the planting moon. The annual report is nearly finished, and soon we must start trekking to all our centres through the area. It will take us most of next month to do so.

Still the war goes on, and daily we look for news of light, but it never comes. The poor Finns have capitulated. What a muddled world.

Tomorrow is a big day in Navrongo, when I must put pen to paper and wind up the financial year. We have about the clumsiest financial system in this Colony that is possible to devise, I should think. Certainly common sense is no guide as to its working..."

From MARJORIE 19th March 1940 Zuarungu "Dear Mother,

Unfortunately, Charles has had to stay in bed this past two weeks owing to an attack of influenza. He felt pretty rotten with a temperature and pains all over him, but I think the doctor will take him off the sick list today. He did try to get up earlier, but the doctor ordered him to stay in bed till his temperature was normal once more. His complaint is that it is almost the first time he has been on the sick list, except when he put his knee out.

We are lucky this March, since the heat is quite bearable still – in the house at any rate. Out of doors it is pretty scorching, and there is a very strong, dry wind. The flamboyant avenue from this house down to the office is in full bloom now, and is really a fine sight, though somewhat startling with its pillar box red flowers…"

The publication by HM Stationery Office of the "Report on Nutrition in the British Empire" by Dr Platt provided full justification for the work we were doing in the Northern Territories. It was so directly in line with the nutritional and economic needs of the people. Kibi Jones, the Chief Commissioner Northern Territories, was transferred from Tamale to important War work on Lord Swinton's Staff at Achimota. George Gibbs came up from Koforidua in the Eastern Province to take over as Chief Commissioner Northern Territories in Tamale.

3rd April 1940 Zuarungu "Dear Mother,

It is almost too hot to write or do anything tonight – one of those steamy nights of which we have about six per year, just before a storm. Very similar to the thundery weather at home only greatly intensified. I find my brain quite addled when it is like this, and my efficiency sinks to a very low level.

Since a little low fever a fortnight ago I have been in good trim. Smith came down from Manga to stay with us on Easter Monday. We trekked into Navrongo on Tuesday, and stayed a couple of nights. But I am afraid we are not settled for long. I am trekking round to see established mixed farmers in this district over the weekend. On the 11th I have to trek to Sinebaga in the south of Bawku district, to look at a proposed land settlement scheme. The idea is to hold back the tsetse fly (sleeping sickness vector) after clearing, with farming. I am going down with Stacy Morris the Medical Entomologist. (168)

Zuarungu is still a dry place, and the white ants seem to eat what the goats leave – between them the garden is reduced to very low ebb at this season. We manage to extract a leaf or two from the garden – rain is the only solution.

Thursday Morning. Alas, no rain. Just after we got to sleep last night a mighty wind got up and the house is covered with dust this morning. There are clouds about so we may be lucky in the next 24 hours. We are fit now, though I have had a slack week with 'flu' or dengue fever. Annual report finished, programme of work for 1940/41 approved. Now we are ready for the farming season. I have now a new man up to assist me – one Smith. Funny that it should be in Wartime our strength should increase. The world is upside down."

From MARJORIE 10th April 1940 Zuarungu "Dear Mother,

It is really too hot to be alive here just now, and I feel like a spot of water sitting on a chair. The average minimum daily temperature is 108° and last month the temperature didn't drop below 79° any night. You can guess that we none of us feel very energetic and yet it is the season when we have to do the most trekking.

We have just heard the news about Norway and Denmark. Pretty bad isn't it?

We have been trekking pretty solidly lately, moving on to a new place every day, and on Friday we are off again to a very distant place. It is famous for bush cow and other game, so I have suggested we do a little tree climbing before starting…"

11th April 1940 Zuarungu "Dear Mother,

A quick note once again to say, amongst other things, that the sky is full of rain clouds and a powerful wind is blowing up once more – excitement grows as we anticipate rain.

These are busy days at a time of year when the climate is most trying. My efficiency drops to a low ebb. Fortunately, Africans don't feel the heat as we do, and the momentum carries the work on over the sticky periods.

Tomorrow, we trek again to Sinebaga where a most interesting experiment may be put in hand. The medical entomologist wants to clear a portion of the river near a cattle and trade route to destroy the breeding places of the tsetse fly vector of trypanosomiasis 'sleeping sickness'. Once cleared, the area of 200 square miles is to be repopulated – if the people will go there – to maintain a tsetse- free area. My part in the scheme is to pronounce upon the suitability of the area for farming. If approved by me, it then becomes an administrative problem, to see if the people will go back.

We are getting news of the latest German invasions as I write and the splendid pasting we are giving the blighters. Life continues very normally here…"

From MARJORIE 24th April 1940 on trek Chana "Dear Mother,

It is 108° in the shade! I only manage to exist by having four shower baths a day, and if the rains don't come soon I fear that there will be a lot of corpses strewn around. It is a very trying time of the year, and one just has to keep going. It is giving Charles prickly heat, poor dear, which must be most irritating.

But he is very cheerful, and all will be well when the rains come. We have had several evening tornadoes lately, but little rain, only enough to make the atmosphere very humid the following day. As Charles said this morning – all he felt like was putting on his sandals and making believe it was a Sunday! But he has a lot of work to do here, and I must say I don't envy him trudging round in this heat. It is just midday.

Since I last wrote, we have been continuing our trekking activities. First of all we had to go to Sinebaga for a few days, for Charles to report on the tsetse clearing scheme. It was lovely country just under the Gambaga scarp, which is like a modified Pennine Range. But large parts of it are now uninhabited, the people having been driven away by the tsetse fly, which give them sleeping sickness. It seems to work in a cycle. A few people die and leave their farms near the river. These get overgrown and the tsetse fly takes possession. Gradually, more people die, and the remainder run away. When the country becomes uninhabited, game comes in increasing numbers. The tsetse flies have plenty of blood on which to feed, and the infected ones increase in leaps and bounds. They go further and further afield, and gradually drive the people a long way away from the rivers. Now there is a scheme, as in other parts of the world, to get people to return.

This is where Chas comes in, for people won't go back unless the land is so good that it pays them to do so. Yet it is useless for Government to clear the riverine areas unless people will farm there, and keep the undergrowth and game back. An interesting problem! Sleeping sickness is curable by injections if taken in time, but they want naturally to remove the cause. We saw nothing more formidable than a snake, but Dr Stacey Morris, the medical entomologist, was spending the following day hunting (a Sunday). He is a dead shot apparently, and in his time has accounted for many lion, bush cow and elephants. I think he got three elephants (169) last tour.

Now we are spending a week around Sandema, Chana, Nakong and Navrongo. This rest house is built on top of a pretty decent size hill, and everything has to be carried up a very rocky ascent. At the bottom a fine new school has just been built, but unfortunately it is situated on very low lying ground, quite unsuitable for a school farm. Charles is out now trying to find the nearest suitable place.

Next month we are to have a lot of folk around. Mr Symond is coming up, then the new Director and then there is to be a Governor's Durbar. How I hope it will be cooler for the latter. Nearly all the officials in the North have to attend, and all the Chiefs and their dependents. Where everybody will stay I cannot think. Certainly every nook and cranny will have an inmate. I expect Zuarungu will be absolutely full although we are 22 miles away. All the Chiefs are received on parade by the Governor and they have a band, races, and plays by school children and so on. It lasts four days.

Last night, we stayed at Sandema and nearly got eaten out of the rest house by white ants. Having retired to bed we heard a chewing noise, and getting out reconnoitring, found that thousands of white ants were emerging from every inch of the walls and floor. They had already started on suitcases, bed bags and so forth. Well, we had to collect every single thing and pile them up on petrol tins…"

From MARJORIE 5th May 1940 Zuarungu "Dear Mother,

We are making a hurried visit to Navrongo tomorrow morning. Actually, we are going to have a few more bugs pumped into us, as prevention against typhoid, just to be on the safe side.

Zuarungu is very full just now for the purpose of a meeting next week; the Governor is holding a Durbar in Navrongo. I in common with many other folk feel somewhat strongly on the subject, as it seems such a waste of money to hold such a thing in wartime when we are being asked to economise right and left. It isn't as if there hadn't been one recently. But there it is. I suppose a time will come when we shall get down to this war business seriously!." (170)

The construction of car ferries (171) across the Volta and Nasia Rivers, at Pwalagu and Nasia, revolutionised life for us in places like Zuarungu and Navrongo. It made access during the rains comparatively easy. Also the toll clerks could obtain quantitative data on the movement of produce into and out of the districts in the North and these, coupled with market data regularly collected in Bolgatanga, gave a very accurate economic picture of the region, of special value in assessing supplies in wartime.

6th May 1940 Zuarungu "Dear Mother,

The season here has been most disappointing. The rains have failed us and planting is already three weeks delayed. Water supplies are low. Out of it all a little good has come, in that I have started a scheme around Zuarungu of encouraging people to dig wells. Don't know why I never thought of it before. The idea is to site a well for a group of compounds. I draw the circumference and lend them picks and shovels, and supervision – the people do the digging. Then I get my mason to cut laterite stone blocks and 'steyn' the well, and debit the cost to the Native Administration. I have three wells being enthusiastically dug at the moment, and many more in prospect. There has been a lot of talk about giving people decent water supplies, but nobody does anything about it. That is the trouble with colonial policy today. There is a rare lot of talk from comfortable armchairs in spacious offices, about what should be done in our Colonies (thanks oddly to Hitler); but a great gap exists between the ideals and the first steps towards them. The first step is generally so simple that it is neglected, and vast schemes never go below skin depth in touching the problems.

I am very busy. This season we have got 18 farmers working with their own bullocks and ploughs, and everyone is watching. Each one is a demonstration farm, and we have got to be certain that they are demonstrating our methods, and making a success of things. My staff is growing at the same rate, and our commitments result in the fact that we are always busy and permanently out of breath. Even this, however, is not all to the bad, because if mixed farming is to go here, it must be in a form which the people are capable of following on their own without a lot of supervision at every stage; so if we make a success of the coming year – and there is every indication that we go from strength to strength – then we know we have started something on a firm foundation which will not collapse when official pressure is removed.

A welcome change is already discernible in the attitude of the Department to North Mamprusi. Now when I write for anything, I get it back by return or a decent letter explaining why not. This month, too, our Director comes up. I gather that he is favourably impressed with work here, having read my bulletin in Kenya, before he came out, and again since his arrival. So if that publication has done nothing else, it has put us and our problems 'on the map'.

I am not expecting great things from the Director's visit, because I am not asking for great things. What I do want is strong confirmation of our policy with a view to the clearing of obstacles – extra-departmentally – for the road ahead when we come to take the next step. Only the Director can do this, as it concerns major lines of policy and development, which are not settled in the district, but at headquarters between Chief Commissioners and the Heads of Departments.

Later – Again mail day overtakes me. Yesterday, we had the second T.A.B. inoculation with the result that we are a little under the weather today. Then this morning I had to address the Mamprusi Chiefs upon the subject of my agricultural estimates. I have asked them for round about £600 this year, which is not a lot for agriculture, but is as much as I can spend properly. The idea is that the central government pays for research, and the local government (the Native Administration as it is erroneously called) pays for extension work – assisted by Grants in Aid from Government if necessary…"

21st May 1940 Zuarungu "Dear Mother,

I am very pleased with a little scheme of mine. We have got eight wells dug or drying since the beginning of the month, and water has been reached in form already not more than 15 ft below ground level. It costs me about £2 to line a well sufficiently for the purpose, and the people are tickled to death with it. They must have a good water supply before they will listen to me about how to assure a food supply. Next dry weather I hope to dig a hundred wells at least.

This lack of rain results in very trying weather. I get impatient and irritable waiting for the rain and the crops we planted on the 10th are wilting. The Durbar in Navrongo last week was a very good show, in spite of the badly chosen time. The natives as well as the Europeans enjoyed it.

This is about the first evening we have had to ourselves for a very long time. We have been inundated with visitors, or else away ourselves. Last Monday, we went to dinner with the Governor in Navrongo! Poor Marjorie as the only lady present had the lion's share of attention, but managed things well enough and got us away early according to instructions. James Broach was up from Tamale for the weekend; the night before we had an old friend of mine to stay, N.A.R. Walker (DC Krachi), and before that J.K.G. Syme. (now DC Salaga and formally of Bawku) Then on Friday morning, the Chief Commissioner called in with the Administrator General from Quagadougou. But it is a pleasure to receive people in Zuarungu when every stick and stone is so familiar…"

In June, we received a visit from our new Director, H.B.Waters, formerly the Deputy Director in the Gold Coast, who had been transferred to Kenya as Director, where he was known as Silent Waters, and thereafter returned to us. I enjoyed his visit immensely; he was knowledgeable and keen on what we were trying to do and supported us with staff and funds, which made a big difference.

4th June 1940 Zuarungu "Dear Mother,

As for farming – the countryside is just grand now. Every shade of green everywhere and crops growing as you look at them. There is some grass now for the stock. Other things being equal, Roland Smith – who has been at Manga – comes down to take over from me here on the 10th, and we go down to Tamale on or about the 24th to relieve James Broach who is due for leave. It is 10 years since I was last stationed in Tamale. A bit galling that now I am going there for a while there has been no polo.

We shall miss the peace of Zuarungu and compare Tamale unfavourably, I'm afraid. However, a change will be no bad thing and a taste of some of the amenities of life – running water, electric light and telephone – will be no bad thing. I shall, of course, not be deserting North Mamprusi; now there is a ferry on Pwalagu, it is easy to get up and down in the rains, and I hope to move about as much as restrictions and economy will allow…"

16th June 1940 Zuarungu "Dear Mother,

Well, at least the much looked for rain has come. The season here is very backward, but it is amazing how quickly plant life perks up when it does come. In some places, early millet is starting to flower – and with it today, a touch of hay fever. The grass land has suffered and with it the condition of livestock. We have had to send most of our stock away from Zuarungu, to better grazing eight miles away. I don't like doing this because we lose the manure…"

From MARJORIE 28th June 1940 "Dear Mother,

Well, leave out here has been stopped, temporarily at any rate; so that Mr Broach will not be leaving Tamale. We spent a couple of days in Gambaga. In the old days long before Tamale was a headquarters station, Gambaga was the headquarters. It was the largest place of the North, and even today it is about the only place shown on the map as far as the NT are concerned. But nowadays it is 30 odd miles off the main road. So life is very peaceful there and one can 'hear' the silence. It lies very high up on the edge of a scarp and the only way of getting down the scarp on the other side is by foot.

Yesterday, we came down to Tamale. We weren't very fortunate as about 30 miles up the road the trailer tyre burst, and we had to leave it on the roadside. Chas has managed to get another wheel and has gone out this morning to retrieve the trailer.

On arriving in Tamale, we found that the Chief Commissioner had very kindly asked that we should stay with him in the Residency. So here we are installed in the very lap of luxury with Vi sprung mattresses, running water and electric light and even hot showers!! Oh, and fresh bacon and butter, the first for six months, aren't being taken lightly. Although we like our bush life, this little holiday is being greatly appreciated.

A few nights ago the Zuarungu staff gave a concert, really intended to be in honour of our leaving for Tamale. It was extraordinarily good. The farm pupils and the overseers took part. They performed in the evening for one and a half hours, by lantern light – mostly action songs and little plays. These people are born actors; completely unselfconscious.

We have had quite a bit of rain, though we are still very short. I have planted up the garden and am hoping for great things in the near future. I have two tubs of moon flowers on the plinth, which are making great headway. We planted 82 potatoes in 'a beautiful manure' (as Mr Yenli says!) but the rains stopped, and I fear that the white ants have eaten all the seed. So yams, I fear, will continue to be the order of the day. They are not bad, though rather stodgy.

We had a rare treat the other day, a piece of pork 'dashed' to us by the

Veterinary Department. I made a couple of quite credible pork pies, which Charles and Roland Smith demolished on top of the Tong Hills after a long ride…"

10th July 1940 Zuarungu "Mother dear,

We are very short of rain. Half the normal so far, and the cattle have been sent away to the river now for grazing. The drought will be very serious if it doesn't break within a few days. The garden is simply hopeless, and we've given up trying with it.

My clerk is away on leave just now, so I am even more busy than usual with the part of my job which I like least. The proposed transfer to Tamale is indefinitely postponed, and Broach has to do a long tour. Now the messenger is waiting to cycle to Navrongo with the mail bag and I mustn't keep him any longer…"

18th July 1940 Bongo. "Dear Mother,

Here we are in Bongo, where I often stayed eight years ago and now have a demonstration farm. The Chief is a charming old rascal, hard, intelligent and very likeable. (172) He has just been up this evening to wish us good night. Yesterday, he appeared with a young sheep, fowls, eggs and guinea corn for the horses, and again today with another lot of stuff. We have not eaten yesterday's sheep yet. But one cannot refuse these tokens of hospitality without giving offence.

These people are really happy under our administration, they develop naturally and freely on their own lines and we can be proud of our achievements. The people are really fond of us, and we are fond of them; there is no hypocrisy in this. The thought of their being otherwise administered is just impossible; in no other part of the world has the administration a higher sense of devotion to duty.

We are still desperately short of rain. The shortage is likely to be serious if we don't have a bit soon; though this particular area is better served than some. The developments of roads, and trade in the last 10 years has, however, made this a far less serious event than it must have been in the old days. It is a crystal clear night and all seems peaceful around. The population is densest here – perhaps 400 per square mile – and all of them farmers; so there is intensive cultivation around compounds which are 50 paces apart. We can teach these people much about manuring, and there is an awakening after eight years.

A drum from the Chief's house and then a bit of shouting; this is being taken up and answered all the way around. It tells that our carriers have been ordered for tomorrow. We are spending a night at Fwegu, about seven miles north of here – on the French territory frontier – where the Chief is likely to become a mixed farmer next year…"

From MARJORIE 15th July 1940 Zuarungu "Dear Mother,

When we arrived here we found Mr Warren was ill once more. He hadn't had anything but toast and water for several days and hadn't sent for a doctor. He's now down in Tamale Hospital with dysentery, a beastly thing to get. He is getting along alright I hear. But he is a terrible person. He is very interested in his work and forgets that there are such things as mealtimes; comes back to lunch at 5.30 pm and so on. Of course, one can't expect boys to keep up to scratch in those circumstances. They get slack as regards everything, flies, boiling and filtering water, and so on. One just can't do those things in the tropics, the price is too high. This is the third time in his first tour he has been sent down to hospital. What does one do with a person like that?

This evening we are going off on trek, first to Bongo by car and then on to Fwegu and Zokko on our horses. Charles has a good demonstration farm at Bongo,"

31st July 1940 Zuarungu "Dear Mother,

At last the rain has come. During the last four days we have had nearly as much as we received during the previous six and a half months. It was just over five inches. We counted ourselves lucky to get back safely from trek yesterday, with the water joined across the road from ditch to ditch, between here and Navrongo in parts of the road.

This is the time to see the farms, so we are trekking a good deal. Having just returned from a week in Navrongo Area, on Saturday we leave for a fortnight trekking in Kusasi (Bawku) and Gambaga..."

30th August 1940 "Dear Mother,

Jimmy Broach is to go on leave on medical grounds, and I am performing the duties of a Chief Agricultural Officer NT in the meantime. But I grow more and more restless, and want to do something a little more directly in line with our immediate future.

In the meantime, the weather is just perfect; grass green and crops growing well. My latest toy is a two bullock mowing machine – the first of its kind in the Gold Coast I should think..."

11th September 1940 Zuarungu "Dear Mother,

We are doing a little trek around the South East corner of my parish on our horses. Crops were so bad a month ago due to drought that it was possible a food shortage might have occurred. But rains have come in ample quantities and the crops have tremendous powers of recovery in the early stages, now everything looks most flourishing.

You will have learnt that we did not get to Tamale after all. Instead we stayed in Zuarungu and I go down for a few days every six weeks...."

From MARJORIE 11th September 1940 Tamale "Dear Mother,

We had a pleasant trek on our horses / bicycle / flat feet and were very lucky to avoid rain, which arrived in the form of a thunderstorm nearly every evening, after we were safely ensconced in a rest house. Charles had a rather nasty cold with a bit of fever hanging around him for about a fortnight. It hasn't been enough to stay in bed, but just enough to make him feel a bit under par. I think possibly he didn't get quite enough leave to remove the bugs from the old system. I am certain that one needs every bit of one's 18- week leave that one usually gets. It takes about half of that to get fit, and the other half to lay in a reserve. He has been taking extra quinine for the cold, which has had the temporary effect of making him somewhat deaf, and in consequence he speaks in a piping voice about four degrees higher than usual!! But I think being out in the air a good deal on trek has done him well, for he seems much better now.

Yes, it is sad about the 'Accra', wasn't it? Thank heavens it didn't happen when I was on her coming out. It was a great pity that the one boat should have capsized – otherwise I think all the passengers and crew would have been saved. We have been remarkably lucky about mail – we seem to have lost very little indeed…"

We arranged to be in Tamale for Christmas 1940 and enjoyed entertaining Hamish Grimm, Roland Smith, Johnny Hinds and the medical officer from Navrongo, Johnny Walker. The harmattan was strong and the weather cool and dry, enabling us to enjoy Christmas dinner in the garden by candlelight. We were delighted to receive a Christmas card from George Cudjoe Chilalah, who had been a pupil at Zuarungu Farm School. He proved so outstanding that he received special promotion, on to the permanent staff of the Department; he also generously donated £5 to the Spitfire fund. We were very heartened at this time by the successes the Greeks were achieving over the Italians, and of course by the calm attitude of the people in Britain to the German air raids.

22nd December 1940 Tamale "Dear Mother,

It is 6.30 am and is as cold as a winter's morning at home – at least it feels like it. The harmattan is a pleasant season for a bit and it is most fitting that it should come at Christmastide and buck us up. The great thing is that there are no insects about. Unfortunately, it is the season of great destruction. Everywhere is as dry as a bone, and bush fires have been raging all round for the last day or so. We too have been doing our bit of burning, but it is of a protective nature.

On Saturday, we returned from a grand week's trekking on our own ground. There is no comparison between the South and North. We are alive and active and going ahead in the North, but here it is clerk- ridden and dull. Our main job was to burn the bush all round Zuarungu district, in order to protect the grasslands between the farms. This is the third year of control and it looks like working well. Anti-erosion is our main reason for control, but there are many other advantages such as better stock feed, better tree growth, higher level of water table through less run off, less flood damage to roads and bridges, and so on. Quite the biggest thing we have attempted in the North and it works in the particular conditions obtaining.

It is all so incredibly ridiculous this war, and could be so regarded if it weren't so serious. How can one tell one's staff to use their common sense, when at the same time Europeans are blowing one another's windows out?.."

Immediately after Christmas, Marjorie and I set off on the grand tour of the west side of the Protectorate, staying with Charlie Orr, Veterinary Officer in Navrongo, and all went to a Hogmanay party with Johnny Walker, the Medical Officer. Thence we went on to Tumu and Lawra, where we enjoyed the hospitality of the Honorary Humphrey Amhurst, the District Commissioner. Lawra district was of special interest because the people, Lobis, who were fairly dense on the ground, lived in scattered compounds. They farmed continuously the land between, like the people around Zuarungu.

We proposed to post an agricultural officer, Johnny Hinds, to carry out an agricultural survey. A special problem presented itself by the depopulation of a riverine area, the Kamba Valley, as a result of sleeping sickness. The Medical Entomologist was engaged in the eradication of the tsetse fly by selective clearing of secondary bush; the big question was, could the cleared bush be held back by repopulation? There were indications that it was soil exhaustion, which caused the first loss of people, and this encouraged secondary bush encroachment. This in turn provided a suitable habitat for the tsetse fly and the harassment of the residual population with trypanosomiasis. In short, sleeping sickness was secondary, and repopulation would be successful only if the land had fully recovered its fertility. There was certainly a need here for a multi-disciplinary investigation with agriculture and forestry taking part.

8th January 1941 at Jirapa, Lawra district "Dear Mother,

It was a pleasant Christmas in Tamale. The harmattan was excellent for the season and provided a real winter's nip in the air, which is a great tonic. A good bonfire gave welcome warmth. We had the dining table on the lawn, a very brown and dried up affair. Boxing Day saw the whole of Tamale civilian population along to a drinks party in the morning, which proved a great success.

On the 2nd January, we broke new ground. I have been down to Lawra and Wa before, but only on very short local leave in 1935. Between Navrongo and Tumu there is much scrubby uninhabited land. Bush meat and tsetse fly have the place to themselves, and there is very little interchange of traffic or ideas. Lawra is a pretty place in attractive country and is the district headquarters. Here we stopped a few days, meeting the District Commissioner, Amhurst and the Medical Entomologist, Dr Stacy Morris. They are most helpful, and thanks to their interest and enthusiasm for agricultural development, I have been able to spend my time profitably. There has been much to do in a rapid survey. The idea is to pick out the main points and problems to have a good start in survey work on this side, similar in method and object to my original survey in Zuarungu.

The population here is not nearly as dense as in Zuarungu, but there is much scope for agricultural improvement. The Chiefs and people are exceedingly keen and they are most interesting. One of the big problems of the place is sleeping sickness (Trypanosomiasis). The trypanosomes are carried by the tsetse fly, which breeds and spreads in the shade round streams and rivers. Morris is cleaning the rivers, and so making much land available for resettlement; it is more than time that our department was doing its job to co- operate.

The people here build the most amazing mud houses like colossal bee hives tier upon tier, so that the final effect resembles a wedding cake. The roofs are flat and beaten, and the run off water is carried away from the walls in wooden spouts. The Chiefs of Nandom and Birifu have three tiers on their houses, each one getting smaller. The top consists of a single room with a flat roof on which one can sit and see the country round for miles. The effect is most Eastern. I have seen nothing else like this on the Eastern side of the Protectorate. I suspect influence from North Africa, perhaps the Songhais who founded Timbuktu. It is an amazing sight to see the scores of people on the parapets – the slave raiders (173) of 50 years ago would find difficulty in approaching these buildings defended by arrows, as they would be seen from afar.

The Lobis are interesting people. They play on xylophones made of strips of wood, mounted in a cradle with calabashes of varying sizes slung underneath. I am told that they have a full octave consisting of six notes.

The effect is most tuneful; with a good camera what abundant material there is for an article in 'the Geographical'.

In farming too, I have been most impressed. We spent last night at Birifu Rest House, planted on the edge of an escarpment looking over miles of country, with the Black Volta running through the middle of the valley.

Here Hinds will probably have his base for the first year. I do not propose to build a house and to make a farm centre until we have carried out a survey and know something of the district and what we want. I envy him his new home. The people have a most interesting sweet potato irrigated garden cultivation, the like of which I have not seen before. It is all commercially fenced too, in a country where enclosure – except for dwelling places – is so rare as to be recorded as non existent.

Tomorrow will be a longish day. We leave Lawra district and go down to Wa for some days. The food and population problems are less acute in Wa than it is here, so that whilst we shall eventually link up the two districts, it will be in Lawra that we shall concentrate at first. Thence on to Bole – a very lonely place, but notable for having Cecil Amory as the Assistant District Commissioner there; we will stay a long weekend and then home direct to Tamale, over the 'Boys Own Road'. The latter I gather is a very bush affair, and very sparsely populated. I hope the old car and trailer continue to run as well as they have done for the first 450 miles of this trek…"

The Lobi Chiefs lived in large multi-storied mud castles. Livestock and granaries occupied the ground floor and mud steps led from one storey to the next. Gutters to take storm water from the flat roofs were made from hollowed tree trunks. The men were good at woodcarving and made three- legged stools with antelope heads carved into the seats. These could be carried over the shoulder by women and used as market stools. The evenings were melodious with the sound of xylophones made from flat wooden bars of varying size, tied to calabashes with leather thongs and mounted in a wooden frame.

From MARJORIE 18th January 1941 "Dear Mother,

We have had to set about getting ready for our grand trek. I've got the art of packing nowadays, but I still dislike it, and I hope in the next life that such things are unknown! It's not only packing one's clothes – one has to pack all one's food as well. Tinned chop always weighs so heavy. But this time we are singularly fortunate – the only thing we left behind was the salt.

Charles contributed to the proceedings by the overhaul of the car and trailer. He hates this kind of work, but the whole outfit was grand – it didn't let us down once, which was wonderful, considering that besides ourselves we had three boys, and Pete and our enormous pig trailer piled high with luggage. Can you imagine it – beds, chairs and tables, lamps, cooking utensils, bath, crockery, food and clothes and thousands of other articles which always seem indispensable.

Well, on the morning of the 30th we set off and spent the first night at Navrongo – that was 120 miles, then on to Tumu. This place has a rather sinister sort of feeling about it. To start with, I don't suppose there are more than half a dozen cars or lorries travelling along the road a year. When we arrived, there were terrific lamentations going on, for the Chief's son had died suddenly – a great pity, for he was one of the few literate boys in those parts. It is a fair sized place, but completely surrounded by bush, and one wouldn't have to go very far to find some excitement in the way of animals. There are supposed to be plenty of lions around there.

Oh, en route to Tumu we passed through Chana and met the Chief on horseback. Telling him we proposed to look in on his farm along the road, he said he would catch us up. In a few minutes he appeared again, chugging along on a motor cycle. Apparently, he had recently acquired this from the Missionaries, and looked terribly pleased with himself. These people take to such things like a duck to water. But he is a particularly go-ahead Chief, and I hear that he attends the little government primary school in his village in order to learn English.

Then on the next day to Lawra, a pleasant place though, of course very dried up now. Lawra and Wa were the only largish places we stayed at. There is a government school at each place and they have a farm attached, upon which which Charles has to advise . The purpose of the trek was that Charles should see the farming as it is at present, and be able to make out a programme of work and select a base for our new man.

I expect Charles has told you about the houses over there, which look very Moorish, being built in tiers. They resemble rabbit warrens, for whenever we appeared on the scene hundreds of people seemed to emerge out of each house. Some of the houses have a juju on top in the form of a devil with horns. The Chief of Lawra presented me with a magnificent bag of embroidered leather. I shall have to cut off a few of the streamers, for it is rather overpowering.

We travelled for part of the trip with the Entomologist, who spends his time in trying to eradicate the tsetse fly from the infested areas. This is mainly on the river banks and water holes. These flies, as you probably know, cause sleeping sickness. This disease is spreading and becoming a menace in the area, so drastic steps have got to be taken. Tsetse flies seem to like damp places which are shaded by certain types of vegetation. So the Entomologist has numerous gangs of boys going round clearing the river banks. They have to be taught, of course, to leave certain useful trees and also not to clear so efficiently that large scale erosion takes place.

It was very interesting as one saw traces of villages near the rivers, which the authorities have proved have been completely decimated by the sickness. This causes the villagers to be gradually driven further away from the rivers. But the infection still goes on, for the people have to go down to the rivers for their water. So the only thing is to clear all the rivers, but the expense is pretty high and is, of course, recurrent. Combined with this the medical people are checking up the infected people in each village. The various Chiefs then have to see that such people attend regularly at the nearest dispensary for their 17 injections. If they don't go regularly they are apparently a menace to the rest of the community; but thankfully the people themselves now seem to recognise the symptoms, and on hospital day, it is quite usual to see between two and three hundred of them waiting for treatment.

It is hoped in time to get the people back and farm the river banks, so keeping down the vegetation. This is where Charles comes in. He has to advise as to what area is suitable for farming. You can guess I have learnt quite a lot on this trek about various soil formations!

Wa, our next base, is a big Mohammedan centre, and at dusk one heard the Amador or muezzin chanting to the faithful. There are many different tribes round here. One of them, the Lobis, has strange customs. The women are duck-billed. When they were children they have a hole made right through their upper and lower lips, and into these holes pieces of grass are inserted. Gradually thicker and thicker pieces are put in, until finally a piece of polished wood the size of a half penny is inserted, (or if they are rich, a coin) in both lips. You can imagine how ugly it is, both lips standing out at right angles to the face. All they could tell us was that their forebears did it, but I think it must date back to the slave raiding days. Perhaps it was done then to make the womenfolk unattractive to the raiders. Around here, instead of carrying their pickins on their backs as is usual, they have a very narrow basket on a leather strap. All you can see of the baby is two tiny feet sticking out.

Our last port of call was Bole, where we stayed with Cecil Amory for a couple of days. He was very fit having just had a spot of local leave. But he has an awful uphill job with the people there – they are very friendly and smile and say yes to everything, but directly one goes away they sit down and do precisely nothing.

Well, so much for our trek…"

On our way down we called at Nadawli, where the Chief was

the father of my head overseer and right hand man in Zuarungu. We had had a hot drive down and as soon as we got to the rest house Marjorie asked Abugari to give her a lemon squash, before he started unpacking. Alas, the bottle that he took from the cold box was a bottle of kerosene. It had been topped up with water from the filter before we left that morning. So great was her thirst that much of the contents of the glass had been drunk before realisation dawned. So we hurried on to Wa and the kind attention of Dr Griffiths, who had a cerebro-spinal meningitis outbreak on his hands at the time.

From Wa to Bole and so along the "Boys Own Road" which ran from Bole to Tamale through Damongo and Yapei. It was badly watered sandstone country with very little population. We covered a total of 840 miles on the round trip and the Ford V.8 with luggage trailer attached behaved splendidly.

My desk was groaning from an accumulation of files on my return but fortunately our clerical staff was extremely good. They were almost entirely men from southern tribes who had been taught good English at school, well trained in typing and office management. Only a minority could take dictation. They would bring up correspondence on the file with a proposed reply in draft, generally in copperplate writing. With the frequency of absence on tour and leave it was essential to keep the paperwork up to date and in order. Financial affairs were important; the individual who overspent a warrant was immediately surcharged.

Towards the end of February, it seemed a good time to take some local leave on the coast.

9th March 1941 Aburi on local leave "Dear Mother,

Local leave was to be the time for tremendous letter writing, yet here we are on the last leg of it and practically no letters written.

However, we have had a grand time and feel infinitely better for the change. It is, indeed, long since I had a holiday, and I felt I needed it so badly...

From MARJORIE 9th March 1941 "Dear Mother,

This holiday has done Charles an amazing amount of good – he stood very much in need of it. The Department has been very decent – Mr. Symond suggested that we should have a few extra days at Aburi, on the way back in order to see it properly. I must say I received the suggestion with great enthusiasm, for even a few days more away from the terrific March heat of the North is a godsend.

When we left Tamale people were just ticking over, and I hear that one of the Europeans there has collapsed and died in our absence. He had a bad heart, but I expect the heat was a great strain. However, we feel fit enough now to cope with most things and we will have the rains to look forward to.

To return to our holiday – not least amongst those who are enjoying it is Pete. He has had an enormous appetite ever since we got to these cooler parts, and is in fine form. We were very lucky in the advice we were given to spend local leave in Ada. It's a tiny fishing place on the sea, just where the River Volta runs out – all sea, sand and coconut palms – not everybody's choice of a holiday, for one sees few other Europeans – but ideal for our purposes and just what we wanted.

There is a very comfortable rest house overlooking the sea, reasonably well furnished and high up from the ground. In addition, there was a fine cool breeze blowing practically all the time. The bathing isn't too good as the surf is so strong, but the entire Coast is like that. However, it is pleasant just to lie on the beach and let the sea come in over one. We had lots of fun with Pete. It was the first time he had seen the sea, and he couldn't make it out at all. I think he thought it was soda water, for he kept rushing in and taking a mouthful. But the crabs were a fine sport.

We were indeed sorry when our holiday by the sea came to an end. These last two days we have been staying with Mr and Mrs Miles in Accra, to enable us to do some very necessary shopping and have the car attended to. Also, Charles was able to visit the dentist, and I the one and only hairdresser on the Gold Coast.

Aburi is really a lovely place 1400 ft above sea level, and at this time of the year pleasantly cool. It is botanical gardens, very like Kew, under the control of our Department. Here everything that might possibly grow in the colony has been tried out – rubber, cocoa, nutmeg, all kinds of citrus, various varieties of palms, coffee and a thousand other things. The lawns are well kept and it is the only place I know out here where one can sit about in the shade during the daytime. There are fine views all round. There is a wonderful avenue of Royal Palms up to this house. We have one more day here, and then we commence to wend our way back to the North. We are a little worried about the car, for the so called Ford agents here have been working on her, and their workmanship leaves much to be desired. Mr Ford would be horrified if he knew the standard. However, I trust that we may reach Kumasi without incident,,,"

A stay in Accra was very reviving to a visitor from the Northern Territories. Cold "chop" from Bikazi or the Ice Company and fresh fruits, papaws, oranges, avocados, mangoes, grapefruit and Cavendish bananas, fresh sea fish and bottled beer made a welcome change in diet. However, the main value of local leave was probably psychological, the relinquishment for a short time of responsibility for a district, a province, a research station or even just the safe keys. There were some who boasted they never took local leave, but often they were most in need of it.

We got back to the North, after 1,100 miles motoring, to the hot season which invariably preceded the rains and made it hard work to wind up the financial year. I had to prepare the annual report but, fortified as we were by a splendid local leave, our task was made easier. Temperatures were up to 110° F in the shade during the day and dropped only to 80° at night.

23rd March 1941 Zuarungu "Dear Folk,

This is ever true of the heat of the last two days. This afternoon was an absolute cracker. I lay on my bed with a bath towel round me and the sweat just rolled off. Now it is lightning all round, and it has all the appearance of a storm in the offing.

Heaven be praised for local leave which has set us up, and enables us to stand this rather trying season. I am up here on my own, having left Marjorie behind in Tamale. We covered just over a thousand miles in our visit to the sea, and were pretty tired of motoring. However, I had to come up to wind up the financial year, and prepare the planting season for the coming year. So up I came with the faithful Abugari on Wednesday evening, and I return on Monday. Fortunately, most of my staff have been working with me for long enough to know my wishes, and I have been extremely pleased with the way the work has gone on in my absence. It is most gratifying and very pleasant when it works like that.

Taking everything into consideration, our work has not been greatly interrupted. It is very important with these communities for known faces to move about in a normal way. Rumours spread like wildfire through the country – the 'white man is leaving us' and things like that. The best answer is to have an agricultural officer going round sighting wells, stopping grass burning, and introducing new and better methods of farming.

However, it is grim to sit and be able to do nothing whilst one's own people are living in hell. This reminds me that leave will be indicated soon. We can be given leave after 20 months though most people have done two years, and some more than that. We are given only eight weeks. However, boats are few and far between to South Africa and people are now encouraged to go home..."

Good Friday April 1941 Zuarungu "Dear Mother,

Here we are back in the bush again. After the amenities of Tamale, I must confess that it does seem at the back of all things. However, with some of our goods unpacked it all looks a bit better. We arrived on Wednesday in appalling heat, and the car ceased to function just two miles from home. It took me three quarters of an hour in the midday sun and a boiling hot engine to get it going again. Today there was a bit of a storm – no rain, but the weather is less oppressive. It has been 110° in the shade maximum, most days this last week, and not been below 80° . At the end of 18 months with the sweat of packing up, one feels the temperature more. I hope I can have some leave in a couple of month's time. Broach should be back on the next boat. A page of grumbles – sorry!

It is wonderful how your letters come through, and how cheerful they are. The horror of it all – perhaps one has too much time to think about it in Zuarungu. At the moment we are without the radio, my batteries have spoiled. It is not good to be without the news. However, if we become desperate we can go up to Nangodi, 12 miles away to Jock Reid..."

28th May 1941 Zuarungu "Dear Mother,

It is another of these oppressive sticky hot afternoons, of which we have had far too many this year. It may bring rain – which is badly needed – but anyway it will bring a storm. Tempers get a bit frayed in this weather, and the body itches from prickly heat, and there is always a good chance of a chill when one's clothes are wringing wet with sweat. Couple all this with the bother of setting up and packing up to go on leave, at the most busy season of the year – if it would only rain! Now you may have some idea of the conditions under which I labour.

Of ourselves there is little to report. We wait for rain and it doesn't come. But we always seem to be doing that. I balance implement loan schemes and settle up office matters, including the preparation of the 1942–43 estimates, although I don't know yet how much I have got to spend this year. The Colonial Service could do with a good overhaul when there is time. It is an incredibly inefficient and ponderous machine at present, and completely clerk ridden."

4th June 1941 Tamale "Dear Mother,

Well we had rain all night – 2.1" fell in an hour on the morning of the 29th and nearly washed us away. Too much rain off the hard ground; but enough was retained to do a power of good.

We broke away from dear old Zuarungu on Sunday 2nd June and spent a couple of nights in 'bush' at Wungu on the way here. Tomorrow off to Ejura, and early on Friday Kumasi. We hope to get a bungalow in Kumasi and to wait there for a boat to Durban. I shall fill in the time writing notes for school teachers, on agriculture in the North.

Having finished the packing and got away, it is a great relief. Already I seem to feel better.

I am sorry that you have not heard from us for a long time. We were notified that about five weeks of mail in January and February was lost.

Great excitement, a few minutes ago a great dog-faced baboon came galloping across the grass in front of the office. Broach shut the door quickly, and it then shot into the clerk's office next door, and out through the window. We were amused to see the staff ejecting themselves from the window in strict order of seniority! Poor brute was lost and nearly exhausted – it could go faster than a man however. Now it has taken refuge in one of our stores on the farm…

This last week has been a very tiring one, for once again we have been packing up, wondering all the time if we should get a wire, as they only give one about 24 hours notice to be on board. Chas has had to hand over to Mr Smith, which entails going around with him in view of all the different activities. He found it most difficult to tear himself away from his two pet dams, which are in the nature of an experiment. We shan't know until we return whether they are a success or not. So you can guess he was in a bit of a whirl, and having got 35 miles down the road suddenly remembered that he had left behind the most important things – Exit Permits and so on, which one now has to have!! Mr Yenli and all the 'chaps' turned out to bid us farewell, and our small boy was on the verge of tears. They are friendly folk.

Just to give us a sendoff, the night or so before we left Zuarungu was pretty noisy. One of our neighbours had died, and every night for about a week thereafter his friends foregather at his house and keep his spirit company. They dance to every sort of drum and native instrument, while the womenfolk shriek and wail. I think about the noisiest instrument is a calabash with stones inside, which they shake. The din is terrific.

The first night it was followed by a first class storm in the early hours – the heavens just opened.

It had been a very hot, dry and tiring spell, and so a cloudburst on practically the last day of the month gives quite an erroneous idea of the spacing of the rainfall. However, thank God, it came then in time to save the crops. The only thing which seems to keep going was the flamboyant trees. To see the setting sun on them when they are covered with bloom is a wonderful sight. When the rain did come, I had to get going on our cassia hedge. I am trying hard this year to fill in all the blanks and make it 'sheep, goat, and donkey proof'.

The goats in particular are an absolute curse. Their owners drive them off their own farms, and let them wander where they will in search of food.

191

They ruin all the trees and shrubs within their reach. And believe me; the reach of a goat after all the most succulent new buds is considerable. I have also put in some mahogany and Neem seedlings against the time when our present trees start to fade away. I find if one does not get this done early in the rains, the time has passed for another year.

As I write, some Fulani herdswomen have come to a nearby well for fresh water. They are certainly very good looking people, with their high-piled hair dressed in the form of a long cone, and their very definite lean features. They carry themselves superbly. They are born herdsmen. A few weeks ago we were in a Fulani Kraal, and one Fulani was carrying a tiny pickin. The freedom from fear with which the baby was bending down to fondle the cows and calves, was a sight worth seeing. Not to mention the delight of the father, in his offspring's tactics. (Our neighbours have just sent me over a calabash of fresh milk.)

Another amusing incident – when it rained the other day I sent down the garden boy for a few trees, one of which proved to be dead. When we were next down at the nursery I, using my best fragment of Hausa, asked the man in charge why he sent me a 'dead tree'; I saw him struggling hard to conceal his mirth and I was wrathful, thinking he was just humbugging me. But it turned out that I had indignantly asked him "Why did you send me a dead man?" the words are very much alike – one is mutum and the other is mutu!

There is a most charming man at our neighbouring gold mine, who is 70+, white- haired but full of energy. Recently he was giving us lots of information about South Africa, when it occurred to someone to enquire how long ago he was there. 35 years come September! He was there at the time of the Jameson Raid. Since then he has travelled practically all over the whole of Africa, south of the Sahara. He was also in parts of Portuguese territory and had, in company with another man, shot many lions in one day. But he is a very modest person and it is very difficult to get him to talk of his experiences. When one can persuade him to talk, he is intensely interesting…"

The Government cancelled leave in Britain except in special cases, and since it was practically impossible for wives to be able to return if they went to the UK we decided it was better to go to South Africa.

We had not long to wait in Kumasi before we were called to Sekondi to join the "El Nil", an Egyptian boat bound for Cape Town via Lagos. Also on board were Isaac and Florence Sibson, of our department. We were some days in Lagos Harbour with rumours of

192

enemy submarines outside. Whilst there we made contact with our old friend Kibi Jones, who had moved there from Achimota. He was still doing special work on the staff of the Resident Minister, Lord Swinton.

The journey to Cape Town took about two weeks; it was a comfortable voyage, and we carried our lifebelts at all times. Fortunately, we did not have to use them, for we noted that the lifeboats were painted in to the davits and it was doubtful if they could have been launched. The cool winter air of the Cape was much to our taste, just what we needed after a tour in West Africa.

Cape Town, set at the foot of Table Mountain, was an impressive sight one could never forget, surely one of the most beautiful cities in the world. Cape Town was buzzing with war activities but not aggressively so. Adderley Street was a centre of smartness, ladies wearing hats and gloves for shopping. At midday the gun fired at the Castle, and two minutes silence followed by the Last Post reminded one that there was a war on.

Our first stop was at Somerset West where Isaac and Florence Sibson had settled comfortably at Mrs Pienaar's guesthouse in the High Street Then we headed over the Hottentots Holland Mountains at Sir Lowry's Pass, through Elgin and Caledon to Albertinia. Here the soil was very red, as ochre was mined nearby. Fine new roads were being constructed with heavy American machinery such as I had not seen before. The roads on which we travelled , except in the large towns, were dusty gravel ones, very corrugated unless they had been recently graded.

We visited the Kruger National Park and Ondersterpoort Veterinary Centre, and I made a point of seeing Dr Curzon of the Royal Veterinary Corps. He and Mr Thornton of Basutoland, were experts on the origins of African cattle, having traced the migration routes from north-east Africa where all originated. I had read their writings, and learned that the ancient British Black cattle had the same origin as our short- horned humpless "muturul" cattle in West Africa. The European cattle went north at Gibraltar, whereas ours followed the coastline round above the tsetse belt.

We visited Basutoland where the most daunting problem was soil erosion, sheet erosion and the most spectacular gully erosion I had ever seen. It was caused by run-off from the mountains, and an increase in population resulting in pressure on the land for farming in the foothills. Cattle dung, which was needed on the land, was used as fuel. All available tree vegetation had long since been cut down, and the problem was to find a tree which would grow there, Robinia pseudocacia seemed promising. The erosion problem of Basutoland was important to South Africa because both the Vaal and Orange Rivers had their origins in the Drakensburg Mountains. South Africa herself was only slowly awakening to the dangers inherent in soil erosion.

Dr Hugh Bennett, the famous soil conservationist, was visiting from the United States. The importance of regional conservation with afforestation, dams and irrigation, contour ridging and contour farming was only just being appreciated. It did not really make much progress until Professor Phillips mounted his outstanding course in Conservation, at Witwatersrand University. He trained some of the first and best conservationists in Africa who transformed Southern Africa. Mr Thornton had a fine photographic record of the work of his department, and was very helpful in explaining what was going on. Unfortunately, my activity in Basutoland was curtailed by a sharp attack of West Coast malarial fever. We made our way back to the Cape through the Free State and the Karoo, spending a night at the Matjiesfontein Hotel which was full of the ghosts of an army headquarters in the Boer War, but the hotel was now in a shocking state of disrepair.

In October 1941, I was offered a passage on a boat called "The Touareg" as far as Lagos. Unfortunately, I had to leave Marjorie behind for the good reason that our first born was due to arrive the following May. Marjorie joined Florence Sibson, who was staying behind for a similar reason. At Lagos, we transhipped to the "Marauder" to go to Accra; she was a very comfortable little boat normally based in Malaya.

Chapter 8 - 7th Tour 1942–43

Trekking around agricultural units alone.

Life back in the Gold Coast on my own seemed very strange. I returned to Zuarungu and buried myself in the work that I loved. I was lucky in that it was just 10 years since we started our survey, and except for a short spell away on the abortive groundnut scheme at Ejura in 1933, I had enjoyed continuity and seen progress.

12th January 1942 Zuarungu "Dear Mother,

I hope you feel no older as a grandmother, relatively, than I do as a father. Actually of course, I'm afraid I have not much idea about it. What I do know is that I feel nowadays I'd like a little more normal company than I get here. (A sad reflection is that I have to be company for myself most of the time.) This Colonial existence is all very well for a younger man, but it is not so good later on. I begin to sense some of the personal problems of the men with whom I once was so impatient and freely criticised.

We had his Honour the Chief Commissioner here last weekend, and he was very anxious to see something of our work as well as the Governor. But everyone is talking about big schemes when a little closer attention to some of the existing smaller ones would be no bad thing. Some of us are never satisfied. As soon as I get over one mountain range there seems to be another. However, we are really making progress in many ways. The grass- burning control has been better than ever this year. I have at last got the Medical Officer to inspect a depopulated area in the east of Zuarungu District. Five dams are under construction at the moment.

Dr Fortes (174) is floating round these parts – you have heard me mention him before. He used to be an anthropologist at Tongo, now out on an intelligence job. I am taking the opportunity of running down to

Tamale with him tomorrow morning, wither he goes to meet the Resident Minister. We spend one night and returned. It will make a break.

The improvement in French relations in Africa has eased things up here quite considerably. It makes the Zuarungu Home Guard seem a bit redundant, which is perhaps as well for we don't want to make fools of ourselves – and I am afraid there was a fair chance of our having done so if anything had happened…"

20th February 1942 Zuarungu "Dear Mother,

I continue my ostrich life – burying my head in work. 'Tis the only thing to do. Of work I have plenty, for I have taken over Dr Cooper's Waterworks Section, while he goes to Accra with sciatica and in all probability home. So I have a tractor making a dam at Detokko as well as half a dozen bullocks with scrapers making a pond; Half a dozen bullocks and about 500 men making a dam for themselves at Shega, eight miles west of here, and all the work mapped out for these gangs, when they finish.

Strange that it should have taken all these years to get going on a water supply for these people, and in the third year of war. Almost as though our consciences were pricking us – as well they might over many, many things. We have just finished a fine dam in Zuarungu. I hope to get three more finished before the rains come – at Bolgatanga, Sekoti and Shega. The natives are becoming dam- minded, and turn out to work on it themselves; we supply supervision, tools and skilled labour. I am looking forward to next season in Zuarungu, for we are at last going to do a bit of straight farming, not experiments.

I keep fit and am enjoying a fine prolonged harmattan season, which has carried on the build up of reserves acquired during my leave in South Africa…"

12th March 1942 Zuarungu "Dear Mother,

I am as usual just overreaching myself with local affairs, and I find it hard to keep track of everything. There is such a big field to cover, and no matter what staff we had, there would always be plenty more to do. Our big job is to keep these people fit, make them self supporting and enable them to produce a surplus for export. This means greater efficiency in production all round, because we provide more of the recruits for the army and labour for the South than anybody. So it makes me weep to see the time wasted at this season by women beside water holes, waiting for them to fill again. So during my morning ride on Humphrey two wells were ordered to be dug – and another cleaned and repaired. Instructions are given to the Chief, that people must cut more grass as bedding for their stock, and ultimately as manure for the land. We have to start such a long way back in production, and it takes time to build up fertility. Laymen seldom appreciate this…"

15th April 1942 Zuarungu "Dear Mother,

Next tomorrow there is another mail out early, and I know I didn't write to you last week because I was on trek in South Mamprusi. At the moment I am at Tono – the farm centre, in the Navrongo Area. I am living in my bush office, while the mason and carpenter finish a new rest house for me. This place is 26 miles from Zuarungu. Things are going ahead pretty fast from this area, largely because of the White Fathers Mission. I have six instructors pushing bicycles round continuously, and a government second division officer in charge of the farm centre.

I am expecting a visit from Guthrie Hall tomorrow, in his official capacity as Acting Chief Commissioner. On Friday, I return to Zuarungu, where His Honour inspects the station and discusses water supplies. On Sunday, Jimmy Broach comes up from Tamale and on Monday we trek together out to Lawra for a few days, to see how young Hinds is getting on before he goes on leave. Did I tell you that, in addition to my own work, I have taken over the water supply section of the Geological Survey, for the time being anyway? This means principally, five gangs of well diggers and a Fordson tractor and scraper, and a dozen bullocks and dam scrapers. It's the very deuce supervising everything.

Smith is away, and I have Manga to look after and all the extension work in Kusasi, as well as Zuarungu and Navrongo. Of course, it does not all get done, but it means that I am never at a loose end. All this is very good, and I seem to think of it as fun. I regret to state I nearly touched 14 stone again. And this, mark you, when I am rushing around the countryside like a mad dog in the hot season, and can get no beer.

Fortunately, the hot season has not been anything like so trying as last year. We have not had our planting rain yet though, and it is always an impatient business waiting for it. It isn't far away. As I write, the Eastern sky is alive with lightning, and now and then a gust of cool air strikes a chill to the sweaty brow.

This is a strange lonely kind of life, where I go for days without seeing a white man, or hearing the news. Sometimes the bush seems so quiet that Pete and I seem to be the only living things on earth.

In spite of myself, I find myself clinging to the old way of living and thinking, and I do feel the need for a salutary reminder. But I must be content..."

Increasingly, with our knowledge of production, market trends and produce prices, we were in a position to advise the authorities on questions of supply and price control. Increasingly, the country became aware of the importance of an agricultural department in time of war. This was further shown by various direct production schemes, initiated by my colleague, James Broach, in Tamale. He developed a herd of 40 breeding sows (a large White Landrace cross) and produced weaners for sending south to the Army Pig Farm at Pokoase, near Accra. This was run by Major Fitt from Rhodesia and Isaac Sibson of our department. The weaners were grown on to bacon size for processing in the Army bacon factory.

A very impressive vegetable garden, covering five acres, was developed in the seepage area below the Tamale Dam for Army and Civilian supply. A poultry unit was also established with improved stock. The oil extraction machinery in the cotton ginnery was put to good use extracting groundnut oil for the army, and the cake was used for stock feed. The hydraulic press was used for baling kapok which was exported for stuffing lifebelts. At last we were not limited by the stage of development of the local farmers, but could go all out with direct production schemes in support of the War emergency. The result was good for our morale and our image. We at last felt we were playing an essential part in the War effort.

6th May 1942 Zuarungu "Dear Mother,

It is 10 years since I first came into this house, on 1st May 1932. There have been many changes in Zuarungu since then, though I doubt if anyone but myself realises them. It has been a fascinating hobby – being reasonably well paid, and being given facilities to preserve grass, plant trees and hedges, make dams, fill drains, sight buildings , sink wells, develop farm schools for training junior staff, start a blacksmith's shop and a carpenter's , running a vegetable garden, to say nothing of the small experimental farms, the cattle, sheep and poultry. I am still continuing to improve the place. My latest craze, if you chose to call it that, is to lay everything out on the contour. The effect is very pleasing to the eye. There is a soundness and rightness about 'fit your farming into the country and don't put square farming into a round country', say the anti- erosion enthusiasts.

So most evenings you will see me out with my level and the six pupils, laying out graded contours. We have done all the arable land, about a hundred acres, and will then start on the gardens. (175). This work has practical applications of course, for we cannot be sure that the protection of grass alone will be effective in combating soil erosion. I have been greatly influenced by what I saw in Basutoland. We will have to work out our own methods for our conditions.

The rains have started fitfully. Whereas we have not been able to plough and plant we have been able to get a move on with our dams. We have finished five since the beginning of February, and it's going to be a tense moment when they all start to fill. I have been working largely from the bush. From this season's work, we should have sufficient experience to make a lot of progress next dry season. I am still leading an ostrich life, but have been so busy, and lately so out of touch, that it is all I can do…"

At this time, I received good news from Marjorie in South Africa that our daughter Sylvia was born on the 22nd May, but her arrival had brought on in Marjorie an acute attack of malaria which at first was undiagnosed. Florence Sibson, too, had given birth to her daughter Gwendoline , and they all continued living in the same household for the next 18 months. Alas, I also had the bad news that Trevor Lloyd Williams (the Beetle), Economic-Botanist in the Department, had died in Kumasi from pneumonia, caused by blackwater fever, and was never to see his own daughter born in the Cape about the same time as mine. Family life suffered.

3rd June 1942 Zuarungu "Dear Mother,

We are getting our rains now, and everything is transformed. It is cool and fresh and the crops are growing under one's eyes. It is a most interesting time of the year. The new contour layouts – which I studied in Basutoland, – are a great success and also our dams are now full of water and have stood the first big test. So literally everything in the garden is lovely.

We are bucket feeding calves for the first time, partly to teach the pupils something about cows and calf rearing, and partly to find the milk yield of the local bovine. The yield is incredibly small. After a bad start the calves are doing well. It is very hard to teach Africans something new, especially if it involves using something with which they are familiar in a different way. They were very sceptical about the beginnings of our self rearing, and prophesied the death of the calves. They would have given up the experiment altogether if I had not insisted. All this means, too, that I get a bit of decent fresh milk for the house which is very welcome these days.

I am enjoying a quiet fortnight in Zuarungu, but start on my travels again next week. Up to Gambaga and then through Bawku district, returning here a fortnight later. It is the best way — the only way to see the country – from the back of a horse. I cover from 10–15 miles a day. My circular tour will take me almost about 150 miles I suppose. Then there is the fun to return to see how everything has grown up in one's absence. It's a good life no mistake, and to be paid to do it too.

Petrol is getting rather short except for special purposes, and since I can afford in my job to go round the countryside more slowly in the rains, I am economising as much as I can. To this end, I bought myself a motor cycle, as a birthday present, and use it for all local running. It goes on top of the trailer when I have to travel, saving petrol on short runs..."

The need to economise in petrol emphasised the value of horse trekking with carriers, which brought one into contact with the people at a pace they could understand. One such memorable trek in 1942 took me to Sekoti, Binaba and Zongoiri, then up the spectacular sandstone escarpment to Gambaga, to inspect the demonstration farm attached to the school. Roy Cooper was now District Commissioner Mamprusi.

21st June 1942 at Manga "Dear Mother,

Two weeks ago I started this trek, by car to Sekoti – 15 miles – and the rest proper trekking with horses and carriers. I went up to Gambaga, three days trek from Sekoti, stopping at the rest houses at Binaba and Zongoiri, a pleasant unhurried existence riding about 10 miles, breakfasting on route. A chat with the Chief and headmen who are waiting; an exchange of courtesies; a talk of the evils of grass burning, the menace of soil erosion, the manufacture of manure, haymaking, the weather, the work of donkeys and bullocks, the war news , increased production and so on.

Then on a conducted tour by the Chief in the morning, followed by an afternoon's sleep or read. In the evening, inspection of my mixed farms, dams, wells, forest reserves; a request to produce 10 carriers at sunrise next morning and another pleasant day has passed. The Chiefs are proud of their rest houses, and like the white man to stay in their towns. It is very desirable to meet them and their people on their own ground, and it all helps in the administration and general advance of the district.

The Forestry Officer trekked with me to Gambaga at my suggestion. This had the double merit of being the best way to exchange views, and also it saved petrol. I pushed on after two days, descending the 2000 ft scarp further along at Saki. I had not been to Saki Rest House since I chased locusts in Mamprusi in 1930, on my first tour. G.F. Mackay was DC, Gambaga at the time, and I reminded the Chief that he and his elders were fined £8 for failing to report on a brown locust swarms to me; this fact was regarded as a bond of friendship between Salmnah and I.

These folk don't mind how odd you may be, so long as they know you are not here today and gone tomorrow; a reasonable enough outlook which Government does not take into sufficient account, when posting staff. I trekked on to Saki; Sinebaga, and Wocambo, where I found Allan Kerr DC Bawku. The views down the escarpment were superb and the air was sweet with the smell of wild jasmine; then on to Garu, where we have a demonstration farm, and so to this pleasant spot Manga. It is very hard on the servants all this trekking, putting up chairs, tables, beds and filters and things and then having to take them all down next day and go on again, so they are getting a good rest here.

I have plenty to do. This morning was good fun, the Bawkunaba – the head Chief here – with a number of his canton Chiefs, came out early on the Native Administration lorry, and I showed them round the farm centre. They seemed to be impressed and enjoying themselves. Roland Smith was away on leave in the UK, and Kusasi District was being well run by George Chilalah. We arranged some pots of 'pito' for them at the end (the local beer).

The Stevenson's screen with the thermometer in it intrigued them, and the rain gauge. I heard afterwards that there is a local legend that the meteorological section is our main rain juju – they see us measuring the rain, and so on. Oddly, Manga has been better served than most places around this season, and the locals have not been slow to explain this. The farm here is really doing very well, it is only four years ago that Marjorie and I trekked past here from Binduri to Bawku, and breakfasted under a fine dawa-dawa tree, which is now in the middle of an experimental unit farm. The land had been farmed out and the people had shifted, and that is why we got 200 acres (176) so close to Bawku town. The recovery of the land speaks for itself and the farmers around know this. Quite the best advertisement for our methods.

On Wednesday, I trek to Zebilla, where the Chief is now working with his own plough and bullocks. My good horse Humphrey has carried me well. Our gentle progress through the countryside must have been similar to travel in Britain in Tudor times. Thursday to Tili and on Friday I pick up the car and go back to Zuarungu. Then it will be time for me to trek Zuarungu district, then Navrongo. In the meantime, I hope my assistant, Roland Smith, will be back from leave.

One keeps fit both mentally and physically on trek. Regular exercise and new surroundings and a complete breakup of routine are very good. Also, it is the only way to gain perspective.

I don't hear the wireless much these days, and feel better without it. It becomes a kind of disease with people. Also, there seldom seems any entertainment. Just talk, talk, talk..."

22nd July 1942 Bongo "Dear Mother,

Dr Fortes bowled into Zuarungu last Monday, also en route to Bawku.

He has lectured at the L.S.E. and Oxford since then, but got so fed up with the smugness and self interest of the latter place – the academic side at least. He got a special political-economic job, and has toured Nigeria before coming here. His company I found as stimulating as usual.

An interesting aspect of life in Zuarungu district, which came to light in our discussions, was the link between the geological formation and the number of generations over which the clans living there could trace back their ancestry. It was naturally much greater on the fertile greenstone than upon the less inherenrly fertile granite sands. Together we could almost construct a geological map of the district. The whole history of people in the savannah zone of West Africa seemed to consist of migration. This is due to soil exhaustion, with the length of occupation dependent upon soil conditions and availability of water supplies. Too often trypanosomiasis, slave raiding or depredations of lion have variously been blamed for depopulation, but generally these are secondary. The fact must be faced that few, if any, indigenous systems of husbandry are capable of maintaining soil fertility indefinitely, certainly not in the face of an increasing population with increased pressure on the land.

This is one of the pleasantest hours of the day. I am sitting outside the rest house feeling comfortable, in a tweed coat and enjoying a whiskey and soda with my 10 grams of quinine. I have become a great cyclist, to reduce petrol consumption... I came along here on my cycle this morning – only 14 miles, but still we are in the tropics even though it is comparatively cool in the rains. This evening I ran up to two places on the frontier, Sambolongo and Bokko where the Chiefs are mixed farmers, a modest distance of five miles away. A nice hot bath and this is bliss.

The faithful Pete has followed the cycle all day – about 28 miles all told, plus all the little excursions. But there are plenty of little pools and streams at this time for him in which to wallow . He loves the water. I have Humphrey here too with me, and shall ride him from here to Zokko and Navrongo across country. He is still going well, but is ageing a bit – and I get no lighter..."

10th July 1942 "Dear Mother,

I am going to try out the new airmail service. I'm afraid there is not one back. Anyway, once one starts on it, it becomes quite expensive at 15 pence a time..."

Roland Smith returned from leave in July 1942 and went through to Manga, thus taking over responsibility for Kusasi which allowed me to concentrate my touring to Zuarungu and Navrongo Districts. From Bongo I travelled west to Zokko, a very densely populated area amongst numerous outcrops of large granite boulders. There was a valuable deposit of clay here and the place was noted for its excellent pottery, cooking and water pots. Also heads for smoking pipes and miniature figures, by a clever man who unfortunately suffered from leprosy. The Chief of Zokko cut an outstanding figure in an admiral's uniform, complete with cocked hat of which he was extremely proud. Then a ride through the bush, on a little- used path for 16 miles, to Navrongo in the next district, then on south six miles across the Chuchilaga River to the Farm Centre at Tono.

Tono (177) was an interesting area; it had clearly been densely populated at one time. Baobab trees marked sites of old compounds which had been vacated long before. Only shallow heaps marked the old walls. The fields had been completely exhausted and the regrowth of vegetation was poor. On sloping land towards the river there were remains of old stone terraces, suggesting long settlement in the past. Incidentally, they were occupied now by countless sand scorpions. I was fortunate in being able to acquire 200 acres for an arable farm and buildings, plus an unlimited area of rough grazing. I soon found that it was asking too much of our system of farming to expect the land quickly to be raised to a high level of production. The whole place was laid out with graded contours and slowly the soils responded; pigeon peas proved to be a useful crop for soil rejuvenation.

To my regret I caused a Baobab tree to be chopped down; it took an inordinate amount of labour, as the wood was spongy. I wanted the site for a rest house, but was told it was unlucky to fell a Baobab. Anyway, I went ahead and built the rest house – a large thatched roundhouse with a wide veranda. It was burnt down the following dry season when the harmattan was blowing, and the locals said "Aha you see". Research in the ruins showed a bottle standing on a window frame on the verandah, through which the magnified rays of the sun could pierce. I had the rest house rebuilt with no more bottles lying about, and no more trouble. During the rebuilding I used the round office as a rest house and, being the dry season, I was sleeping out. I was very impressed to see the masons, carpenters and labourers going to work in the early morning reverently doffing their hats and passing silently by in single file.

Sunkari Wala was my head overseer here, a product of the training centre, Cadbury Hall in Kumasi. He was a man with a good sense of humour and a high standard of responsibility. I was well served by my staff; indeed anything I was able to achieve was due to them. They gave me good local counsel and were keen and trustworthy.

So back to Zuarungu where it was astonishing to find how the paperwork accumulated, after an absence of only 10 days. I was beginning to feel that Government was taking too much for granted. There seemed to be a feeling that because there was now an agricultural presence in Zuarungu, all was well and there would be no famine.

In fact, because the economic tempo had increased, partly due to an economic awakening as a result of agricultural development, there was a growing need for better road facilities. A good map of the area was needed; telephone and postal facilities were essential and a good senior administrative officer would enable a more efficient agriculture to function effectively.

I was pleased to learn at this time that the Secretary of State for the Colonies had agreed to my promotion to Senior Agricultural Officer, which meant I received a seniority allowance of £72 a year when on duty.

7th October 1942 Tono Rest House "Dear Mother,

I am on trek as you will see, and staying at the rest house I built on our farm centre here earlier in the year. It is a pleasant little house – a large roundhouse with mud walls and a tall conical thatch roof, concrete floor, with small bathroom, pantry and veranda attached. Kitchen and boys rooms are separate little roundhouses. The whole is perched on a little hillock, with the farm in the foreground between me and the main road, half a mile away. In the far distance are the hills in French territory, and in middle distance, five miles away, the Navrongo ridge.

It is so nice having one's own place, instead of living cooped up in a noisy rest house in Navrongo – an ugly place at the best of times. We are busy mowing hay, got the only mower in the Gold Coast here, a little one horsepower mower, a Bamford, pulled by a pair of bullocks. It is a most homely sight and sound, and saves a lot of labour. I got it four years ago, and got into a spot of bother for ordering from a local firm instead of through the Crown Agents…"

16th October 1942 on Trek Sambologo "Dear Mother,

I rode in here, a three hour ride on Humphrey, and write a few lines for my messenger to take back for the post early tomorrow morning. I expect him here with the mail any time now.

Left Zuarungu yesterday after tea for Gawri, tomorrow I go on to Bongo, and return home on Saturday. On Monday I start off in an easterly direction for Manga – the rivers cannot yet be crossed so that will be a two day trek of 45 miles. On Thursday, we are having a district conference, to discuss the work for the dry season and particularly the programme for grass burning control. It seems no time since we did it last year, just after my return from leave.

My present trek is to site places for dams, and to put a level over them so that I can calculate the best site and the size of the bank. So you will see that in spite of everything I had an active and interesting life. It is a good thing not to have much time to brood over things these days.

As I rode along early this morning the blue sky was a mass of fine white clouds, which softened the glare of the sun, and as I look from Gawri towards the rocky hills of Zokko I could almost believe I was riding along at Santon. It looked like the scene above the road where I often walked. Of course, there was no swishing sound and the sea on my left, so I cantered along for three miles, where the road was good, in a blissful atmosphere of sentiment. Funny how just a particular view or landscape can remind one so fairly of another place, another time. But soon the sun got up, and when I crossed the rocky horizon, the path deteriorated considerably. For the rest of my ride it was just a mass of granite boulders with poor parched eroded groundnut farms in between. I was glad to get to the rest house here, some minutes after twelve. So was Humphrey, I guess. He is getting old and I grow no lighter.

My carriers were supposed to be cutting across on a more direct route, and I was surprised not to find them in when I arrived. It was some minutes later when I heard Pete panting outside. The Chief of Sambolonga produced a magnificent deckchair however, in which I am reclining now, and enjoying a welcome glass of ale. (We are fortunate in having a brewery in Accra which turns out quite a potable product.) Last Friday, I found I had a bit of extra petrol left, and so I went south to see the demonstration farm at Binski and then up the scarp to Gambaga (1700 ft) to see how the school farm was getting on. Both farms interest me very much. Conditions in South Mamprusi are so different from here in the North. Gambaga is such a restful place, with plenty of trees and bird life. I met the new Provincial Inspector of schools, one Dickens, with whom I found I had much in common in general outlook.

My heart has thwarted my head and I have decided to form a Home Guard unit in Zuarungu, composed mostly of my staff and the school master. There is local enthusiasm amongst Africans and I cannot any longer hold back, if only to encourage them. Now the crop season is finishing I shall have more time to devote to it, and I intend to make it effective as far as I can in the circumstances.

Back in North Mamprusi, the new dams were holding up well and recruits continued to come into the farm centres for training in mixed farming. The supply of ploughs from Ransomes, Simms & Jeffreys of Ipswich continued to be available in spite of the War. For Christmas Roland Smith came down from Manga and together we went through to Navrongo and stayed with Arthur Russell, the Forestry Officer. It was a jolly occasion, no less than 27 enjoyed Christmas dinner with Ted and Bun Ellison. Bun was the only lady present, and our numbers were swelled by a recruiting party of three from the Regiment in Tamale. There was a general feeling that the tide of the war had turned, and the news was certainly more encouraging; on Boxing Day Roland Smith and I went out to the rest house on the farm centre at Tono for a couple of days, then he returned to Manga and I to Zuarungu with money from the Treasury to pay wages. After the payout I rode down to Tongo to see the New Year in at the rest house at the foot of the Tong Hills, then a riding tour to Shega and Zanlerigu to look for dam sites…"

20th November 1942 Zuarungu "Dear Mother,

A thing which is bothering me now is the rate at which the locals are ruining their land. It requires a minimum of 10 years in a place to learn everything about land utility. I doubt if there will be a farmer living here in a hundred years time, at the present slow rate of our extension work, and the present rapid rate of soil erosion deterioration. I have been investigating the problems this month.

One of depopulation of a large riverine area strip, associated with blindness and skin disease. (Deficiency disease) I am trying to get the Medical Officer to go to visit the place, and find out what the remnants of the population are like. Anyway, the soil is completely exhausted in these places. The other problem is overstocking and soil erosion on the high land around Bongo, which is one of the places to which the people have crowded as a result of 'bad farms' and the 'land becoming bad'. Government is apathetic. I want roads through to all these places, so that we can see them more easily, and a team of people to work on the problem. Of course, wartime is bad time to be asking for these things, and I'd be satisfied with the assurances that the problem would receive attention when the fighting stops.

But Government's attitude was the same before the war, although 10 years ago it was concerned and that is why I was sent here. I would feel that I had wasted much time if I didn't get something done before I left the place. And I cannot expect to be here always. An idealist you see, mother dear. I think I must have got that from you, and it is proving an awful nuisance to the administration. Now I promised I'd go early to the farm this morning to fix up the new cream separator, forgetting it was mail day…"

I was becoming obsessed with two fundamental aspects of life locally, the first was water supplies and the second soil fertility, and both were concerned with the location of people. Right across the north of the Gold Coast from Bawku in the east to Lawra in the west the riverine areas which had once been populated were now ruined by soil exhaustion. The question was where the population would go when they had exhausted the land they were now occupying. Could we with "mixed farming",(178) bushfire control, contour ridging, strip cropping and the construction of wells and dams significantly turn the tide? I was worried because I felt the administration tried to evaluate the policy of one department against another – veterinary, agriculture, medical, education – instead of getting them all together to contribute to a single Government Policy which all had to support. We were getting near to it before the War, with Kibi Jones's annual conference of Northern Territories' Officers, but it needed to go further.

23rd November 1942 Zuarungu "Mother dear,

It is just after 6 am and the world has a rosy colour as the sun rises through the harmattan haze. I am sitting out on my stoop enjoying my early morning tea. Now that the mason has come, there is an inward urge to be up and doing. That is if there is work to be done in the open.

I have had a very busy day today, taking a surveying class before breakfast. Met the fire sub- divisional chaps to discuss a day for burning the bush in this area, after that the transport men in Bolgatanga, to discuss marketing poultry – or rather the times of closing the ferries further south. So I must hurry and shave, mother dear, and get the day started.

The weather has become trying here these last few days, and the harmattan has left us temporarily. There has been a good deal of electricity about, which I find affects me quite a lot. It has really been nasty and tropical, and I have yearned for a deep cool breath of the English Countryside. But when I shall draw it again is another matter, which concerns me just now. I cannot see the war ending before I am due for leave – and even should it end quickly, it would take some time for Marjorie to get a passage home. I doubt if she would get one home now, even if onc said to blazes with the submarines…"

9th December 1942 "Dear Mother,

I am sitting on my stoep watching the sun rise and the warm glow on the Tong Hills, as I sip my morning tea. The scene is a familiar one of which I have never tired. But alas, the time is coming when I shall have to leave it. Nearly 11 years in one station is an all time high Gold Coast record for a District Agricultural Officer. But I learn that the Director is contemplating moving either Broach or myself, since we are both Senior Agricultural Superintendents, and in Ashanti and Central Province there is nobody of that rank.

I will trek for two days to Binshi today. Binshi is down over the river in Gambaga district. The instructor there has built me a grass shelter, for the vast sum of 30s since there is no rest house at hand. So I shall be amongst the trees tonight all right, and Abugari will be cooking on a stone in the yard…"

10th December 1942 Binshi "Dear Mother,

I certainly was amongst the trees last night, at least until 1.30 when I was rudely awakened by a loud clap of thunder over my head, and only just got my bed moved into the tiny little grass hut in time. Quite wrong of it to rain at this season! There wasn't much, and the air has now become a good deal clearer today. As I sip my tea early this morning an enormous bomber sailed overhead. We don't see many planes in these parts; and it's quite exciting, wondering where it has come from and wither bound. In these days, with the French Colonial situation somewhat in the balance, one gets the feeling that 'things' are going on all round one. The news sounds hopeful, and I feel sure that time will sort the situation out. I do agree with De Gaulle that there can only be two kinds of Frenchmen – those with us and those against.

It is amazing what a change coming 30 miles south from Zuarungu makes. Here on the sandstone, the people live quite differently, look different, and grow different crops and have different problems. A remarkable change not sufficiently recognised by Government. The trees and grasses are different in many cases, and water is a problem – much more so than in the North. Possibly because of this, the population here is one tenth as dense as in Zuarungu. The importance of South Mamprusi is as a reception area from the overcrowded North, when the people there spill over, as most certainly they must sooner or later.

Our problem is to ascertain in advance of the panic, the best way in which this area can be developed. Then we can give sound advice to immigrants, and control settlement. My object is to carry out an agricultural survey as soon as possible, to find out all we can concerning the people living here now. Then put in a small experimental farm, in the meantime, training staff, for the extension of results in the future.

I wish I could share with you some of our cream. I got the separator which has been lying in Tamale store for many years and never used. We make half a pint of cream in the morning and sell a gallon and a half of whole milk, minute from seven cows, but a start, and a new thing in this country. Our cows give an average of 120–150 gallons in a lactation period of 300 days. We have raised all our nine calves by hand with success, and the pupils have learnt much thereupon…"

19th January 1943 on trek at Shega "Dear Mother,

There are many trees and shrubs flowering at the moment, and it is surprisingly sweet cantering along on Humphrey in the early morning – surprising because Africa is full of smells, most of them anything but sweet. Wild gardenias and acacias are chiefly responsible. I am really looking into the reasons for population movement which have taken place since the 1931 census.

My work is too often regarded as a function of the Department of Agriculture, and not as a function of the Government service as a whole. Thus, the administration is apt to try to weigh what they consider to be the 'rival' policies of 'rival departments', instead of co-ordinating the work of a number of complementary departments. It is in matters of this sort that the advantages of continuity are so great. I have put up a scheme for Land Planning and Rural Development committees in districts, provinces, and heads of departments in Accra, which has, I believe, gone up to the Governor. His Excellency is coming round next month, and I hope to have a chance of crystallising the subject. The idea could work if the will to make it work existed at the top. I'm an idealist, as you know mother dear.

There is too much talk, talk, talk these days about specious schemes, and not enough attention is paid to the simple, fundamental things. I am becoming an old hand at bringing these matters down to water supplies or soil erosion. Two matters which, above all, are important in Africa today. Too many people sit in offices, and too few people get out into the country and do a job of work. It is not surprising that it is amongst the farmer that one finds most dissatisfaction, and eternal discussions about rights and seniority.

Ah, I see the signs of lunch approaching – bush guinea fowl, strongly recommended today's dish…"

28th January 1943 Navrongo "Dear Mother,

It is still cold, dry and harsh. I am doing a valuation to the aerodrome, and living on the spot, and it is very exposed to the winds that blow. It is an odd sensation to be in temps of 65 to 105° , with a perfectly dry skin and cracked lips, a little trying after a bit. I feel I would like to have a good sweat. All too soon, however, will 'lokachin Zufa' (the time or season of sweat) descend upon us and we shall be grumbling about something else.

Actually, things are going along very nicely. I have had considerable encouragement of late in the extension of our ideas in Zuarungu area, and am having the satisfaction of seeing quite a reasonable proportion of my dreams come true. This is as much as any of us have a right to expect…"

In February 1943, we had a visit from His Excellency the Governor, Sir Alan Burns. (179) Neil Dobbs, our new Assistant District Commissioner, and I decided he would like to see something of the depressed riverine areas to the north of Nangodi, towards Arabe and Fwegu. We hastily constructed a bush track with railway lines chained loosely together and pulled by the Fordson tractor of the Water Supply unit. The plot succeeded in spite of a broken spring on the gubernatorial Lancaster. Sir Alan was interested and promised to try to send up an army photographer. What we really wanted and got was of course his support.

He was as good as his word and in due course Lieutenant Clements arrived in a small army van all the way from Army Headquarters in Accra, murmuring that Generals should not dine with Governors. He was even more dismayed when we showed him what he had come to photograph, but he warmed to his subject and after a few days became very keen and took around 40 wonderful photographs. Some time later, we received an album of these photographs setting out the problem of soil erosion in the Northern Territories, which we used to good advantage; it provided a permanent record.

14th February 1943 Winneba by the sea "Dear Mother,

I had a sudden urge to get away from the dust and dryness of the North, and to see the sea. Our department is always good at granting local leave; of course, they know we would not ask for it if we felt we could not take it. So I came down from the North as soon as the Governor's inspection was over. A weekend in Tamale with James Broach, a night at Ejura and so to Kumasi for a single night. Hence by train to Nsawan. Very odd catching a train again after 14 months. A night with Sibson under canvas on the Accra plain, about nine miles from the big city itself, then a lorry from Nsawam to Swedru.

Winneba is a lovely little station, I have never been here before, but have made up my mind that this is a place for local leave in the future. The bungalows are perched up half a mile back from the sea, dotted almost on the edge of a little ridge. Around the residential area there is quite a fair turf and quite the nicest golf course I have struck out here.

The sea is nice for bathing and the grass leading down to the sand is just like the downs in miniature. The rest house has running water, electric light, and diffusion radio layout. Quite ridiculous how the money has been splashed about down here – of course this is where the money is to be found, or rather just north of here in the forest. Winneba was once an important cocoa export port…"

On my return to the North I packed up in Zuarungu and came down to take over Tamale station. Tamale is an interesting place just now, and it is time I was shifted from Zuarungu. I was glad to be able to show His Excellency round before I left. Actually, he is not very agriculturally minded, but he tried, and I think I got a few points across.

16th February, 1943 "Dear Mother,

I am in a bit of a quandary about my next leave. I would like very much indeed to come home; the problem is what to do with a wife and child in South Africa. It is a funny sort of life this really, when for 18 months I might just have been reading about a family in a book, for all I know of it— or for all that it enters into my life, apart from my bank statement. Marjorie suggests leaving Sylvia with Mrs Pienaar (180) for six months. She could get a priority passage and a bit of extra money by going into the Chief Commissioner's Office in Tamale, to help with the confidential stuff which cannot be left to African clerks. But I don't want to place her in the invidious position where she has to choose between child or husband. I can look after myself better than Sylvia, but on the other hand this kind of married life is just not married life at all…"

It was not an ideal time to take over, with the closure of the financial year and the annual report due at the end of the month. Delay would have meant extending the length of my own tour, which was not sensible with the pressure under which we were now working . Again I enjoyed Tamale with its amenities; we were doing so much by way of direct production for military and civilian supply. Also buying grain, guinea corn, millet and rice for army consumption, so that at last I felt we were pulling our weight in the War effort. One reflected on the length of time it took to mobilise all forces. Tamale carried a breeding herd of 60 sows, 100 cattle, 100 sheep and 60 poultry, the latter supplying eight dozen fresh eggs daily to the Army Supply officer. In addition, there was fresh produce from the five-acre vegetable garden.

23rd April 1943 Gambaga "Dear Mother,

Here I am in old Gambaga. It is 'old' in more ways than one; old to me because here I did my first trekking, when chasing the wily locusts in my first tour; and old because it was one of the earliest stations in this part of the world. But gone are its former glories. The Colonel's house, on the hill at the back, has not been lived in for 30 years or more. The kola and salt trade-route to the Haute Volta and Niger regions no longer passes this way, but by lorry up the main road 30 odd miles west to Bolgatanga. But it all serves to emphasise the peacefulness of the place. The station layout and the trees somewhat overgrown. Gambaga is higher than some other places in the NT being some 1600 ft above sea level as against Tamale 600–700 ft. Perhaps this has something to do with my feeling of wellbeing.

But I have not come here entirely to play. We have a very interesting demonstration farm attached to the Native Administration School here, which I have neglected since last August. Also, I am hoping to get a lot of Kapok floss (185) from here, and am arranging to send up a man to organise a purchasing and cleaning scheme. Our job is much more interesting these days because we really are at last being used for war work, and have funds available to get on with things.

I went up to Zuarungu last Sunday, picked up Roland Smith and went across to Lawra to see Johnny Hinds, (186) and to see how he is getting along with the new farm centre we are starting up across there. It was a pleasant and uneventful trip, and quite successful from all points of view. These two lads, Smith and Hinds, are doing a dam good job of work; then a run up to Bawku on Saturday, and round here yesterday.

It was strange visiting Zuarungu again with somebody else living there. But I felt no regrets about leaving, rather a relief, to have the detail of district work off my hands. My present job, which carries the lordly title of 'Chief Agricultural Officer, Northern Territories'" gives me the scope I want and feel I can use, except that I want a large staff of Europeans up here instead of five. Won't get them yet, but there is work enough for them.

Tomorrow morning I go out to see the paramount Mamprusi Chief for a pep talk on production. Thence to Tamale, where I've no doubt the work has been piling up during my absence"

We were much encouraged by the number of administrative officers who came to stay in our rest house, and learn something of our policy and philosophy. The Chief Commissioner often rode out with visitors to see the Tamale Experimental Farm in the evening. The swimming pool in the old cooling tank by the ginnery proved quite an attraction, particularly to officers in the Regiment. They were pleasant days indeed.

14th May 1943 Tamale "Dear Mother,
I continue to enjoy Tamale, but the weather has been very trying these

last few days, with storms hanging around which never break, and I am feeling tired. On such occasions, I wonder why I ever came to the tropics, and yearn for England's green and pleasant land.

The news is good these days. Those chaps up north made a first class job of it whilst they were about it. And the news is of mighty blows by the Air Force on Germany herself, though losses are tragically high.

My large vegetable garden is having a bad patch just now, and the army supply officer has had to fall back on tins. It is so hard to give personal supervision to every detail in the garden, and yet without it very indifferent results are achieved. However, things should buck up after the rains…"

17th May, 1943 "Mother dear,

Colonel Gibbs asked me to come up to see him this evening, about a matter he could not discuss on the telephone – judge of my surprise and feelings, mother dear, when he suggested a stroll in the gardens and handed me a letter to read. I thought it might have had a petition from one of my staff. But it was marked 'Confidential' and bore the headings of Government House. I read it in a sort of daze but in strict confidence (and dam the censor); (183) it was a letter signed by Sir Alan Burns, asking the Chief Commissioner to give me his congratulations and asking if I would accept the honour of Member of the British Empire. I felt completely shattered, and rather like a little boy climbing onto your knee to tell you a secret.

Needless to say, I said I would accept the honour, feeling that it was for you, mother dear, to whom I owe everything. How these things are awarded I cannot think, but I do feel terribly unworthy in myself, and rather accept it in memory of you and the numerous outstanding people it has been my privilege to know and love.

I feel bewildered. However, not a word. I presume it will appear in His Majesty's Birthday Honours List…"

In the King's Birthday Honours I duly received my award for services to agriculture. I showed the Governor around on 19th July. We had a success story to tell, – with a staff of two agricultural officers, eight Government assistants, nine local government assistants and 16 itinerant demonstrators. We had built three farm centres, 16 demonstration farms, nine demonstration dry season vegetable and fruit gardens and established 85 plough farmers.

Twelve dams and weirs had been built, and dry season grass burning had been dropped completely in Zuarungu district. This covered an area of over 2 million acres, with a population approaching half a million people. At a garden party that evening at the Residency, Sir Alan Burns invested me as a Member of the Order of the British Empire. My old friend, Major Owen Butler, Acting Chief Commissioner NTs read the citation. Kibi Jones received a well deserved Knighthood in the same honours list.

"9th June, 1934 "Dear Mother,

I am feeling pretty fit, but tired. There is no letting up in this job. So much routine which takes longer when one is tired, and leaves no time for thought about policy and the larger things. But this is entirely due to the extra work in connection with the war, which I am very pleased to do."

"15th June 1934 "Dear Mother,

My mail bag is a big one these days. Surveying the wires, letters and handshakes I have received on account of my Birthday Honours. Africans, of course, simply love it, and if my own staff takes the pride in it that I take in them, we shall all be happy. I think people are pleased particularly that somebody living away quietly in the bush can be recognised in this way. Most honours go to people in Accra."

Life in Tamale was pleasant and rewarding, particularly in the rainy season, when crops were growing and the air was comparatively fresh and cool. There were plenty of worthwhile things to do, with army supplies taking up most of the time. The local mail came north from Kumasi on a Thursday of each week, and went south again the following day. It was a great mistake, I found, to try to reply by return of post. So often the information had already been given to Accra in a quarterly or annual report.

I heard in June that Marjorie had left our daughter, then 13 months old, with Mrs Pienaar in Somerset West and sailed from South Africa;(184) I met her from the train in Kumasi and we came north straight away because a Rubber Commission was taking up all the available accommodation at the time. Marjorie worked in the District Commissioner's Office in Tamale on petrol and kerosene control and was also a member of the W.V.S.

"14th July 1943 "Mother dear,

I am sorry you had no letter from me. I can assure you that it is only the fact that I seem to be so busy these days that it should be so. All day long the telephone rings in my office. Then this place is full of committees, and there is not much that does not affect agriculture in some way or another.

On top of all the extra war work – kapok buying and pressing, groundnut oil extraction, pig and poultry breeding, butter making, and army produce buying – the District Administration work has to go on, and the farm calls for attention.

I must say that I do appreciate having my house run for me and a bit of comfort in my old age. With having to do a long tour it makes quite a difference. From feeling a bit low some months ago, I gain strength month by month."

4th August 1943 "Dear Mother,

The amount of paperwork in this job is just ridiculous. If Head Office calls for a single return more it will be the last straw, and so much time will be spent in the office that there will be no fieldwork and nothing to report; presumably then we start all over again. The trouble is that there are too many Directors of Agriculture in Accra – one chap deals with Army supplies, another with special food production, another with rubber, and so on, and they all sign for Director and don't realise that all letters converge on one wretched bloke somewhere who has a large farm, a large vegetable garden and an enormous territory to run as well as an office.

We are having quite a serious drought in this belt. But the weather is clear and cold and very healthy."

About this time mepacrine was introduced to replace quinine as a prophylactic for malaria; most people had reacted well to it but tended to go yellow in colour.

22nd October 1943 "Dear Mother,

It seems to be strange, in a way, to be visiting my old spiritual home in Zuarungu. This is the first time I have been back since I left last April. I feel as if I had put my life into reverse. The experience is not unpleasant. I am simply confirming in my own mind the solid pleasure I received from very simple things here. Yet I feel surprisingly few regrets at leaving, and have no great desire to return to live here.

Unlike Dagomba district, of which Tamale is the centre, this part of the world has had fine rains. Indeed, last Thursday, the day we arrived, the rainfall reading was 1.75, the day previously it was 3.75 and on Friday 3.1". Everything was pretty wet as you can imagine and the road to Navrongo was breached in three places. We took the precaution of putting the car and trailer seven miles down the road on the other side of the Winkogo embankment which sometimes goes after prolonged rain. The countryside here is lovely, thick green grass and fine crops. It has been interesting too, seeing our anti-erosion works of last year standing up well to the rain. All the dams are full and brimming. Some are very good, others not so good, – and one has breached, but it was all worthwhile. Now the water supply section is being started up again, which will relieve us of this extra work.

I often yearn to do farm work and don't mean sitting on a tractor all day – One misses doing the job for oneself in this country."

At the beginning of October, Jimmy Broach returned safely from leave in the UK and after handing-over, Marjorie and I made our way to Kumasi to be ready for the next boat to South Africa. It was restful in Kumasi and cooler than the North.

29th October 1943 "Dear Mother,
Kumasi is disgustingly well stocked with all kinds of materials and Marjorie has been lucky in borrowing a sewing machine. She has been very busy making clothes for Sylvia and herself. People seem to be short of nothing. Fruit is abundant and cheap here, but meat is not as good as in the North and there are no horses because of the tsetse fly."

On the 16th November, we left Accra for Winneba where I had enjoyed local leave earlier in the year, we were lucky to find a house available and it proved a most agreeable place in which to loiter. We were there for nearly a month, and both went down with malaria fever just at about the time I had completed a two year tour.

In mid-December, we were called to Sekondi because a boat to South Africa was imminent. On December 15th, we were whisked away in great comfort on the M.V. "Rimutaka" of the New Zealand Line; she was a fine ship taking only nine days to reach Cape Town on Christmas Eve. We went straightway out to Somerset West, I to meet for the first time my daughter who was by now 19 months old and just recovering from whooping cough.

Chapter 9 - 8th Tour 1944–1946

Agriculture in the War, Acting D.C., Family Reunion.

After many happy months at Hermanus my leave expired and we decided to move closer to Cape Town, to be ready for a boat back to the Gold Coast.

> 20th January 1944"Dear Mother,
>
> These short leaves are not good and I certainly don't feel fit again yet. Whether or not Marjorie returns with me depends upon factors outside my control. If perchance we should be posted to Accra – and there is just a remote possibility – we could all return to the Coast – three of us –and Marjorie could bring Sylvia home after a few months. But so far as I know at the moment I shall be returning to the North again. I do not think Sylvia should be left again on her own, so Marjorie would have to stay down here, and in due course endeavour to get home. It is now that the disadvantages of service in West Africa are most pronounced for a married man. Apart from separation from one's family, the expense is appreciable."

Finally in mid June 1944, I secured a passage on the "Cairo City". My wife and daughter went to Katberg in Cape Province to stay in the mountains. We spent many nights on deck because of the smell and lack of air in our cabin, normally a two berth, which we shared with four others. Investigation showed that some of our trouble was due to a large over- ripe Roquefort cheese. It was wrapped in shirts and socks and stuffed into a locker belonging to a colourful professional footballer, who occupied one of the upper bunks. The culprit then remembered another similar cheese packed with his dinner jacket in the baggage room!

9th July, 1944 "Mother dear,

I landed at Accra three days ago after a somewhat uncomfortable, but uneventful, journey. Now I am getting my car on the road again and collecting up my goods and chattels, ordering stores – a job I find increasingly difficult. Zuarungu calls me and I am leaving as soon as possible for the North. Rains have failed up there and there is, or has been, a period of food shortage.

Now locusts are threatened so it looks as if there will be a job of work to do. The Governor is also going up next week. I have been away so long that I will not have much time to catch up with current matters.

I am missing the family very much. It is a hopeless way to live maintaining two homes. So far the Government has in no way recognised that prices have risen from 100% to 200%. Fortunately, I am well up the salary scale now. How junior officers manage I cannot think.

The faithful Abugari was on Kumasi Railway Station to meet me. Just how he gets to know I was coming is a mystery, but I was darn pleased to see him as you can imagine. He immediately took over as if I had been away only a couple of days.

We are stretched to the limit. It was Rhodes I think who said 'too much to do, so little done'. The Agricultural Officer for the Mamprusi-Navrongo District might well adopt it as a slogan. During my absence our title has been changed to Agricultural Officer from Agricultural Superintendent, a much better nomenclature. I never could understand the Superintendent business – always seemed to imply something to do with sanitation. But what's in a name!

I have been writing this sitting on the stoep as I sip my morning coffee. The sun has risen as I have been here and I must get on with the day."

1st August, 1944 Zuarungu "Dear Mother,

I have been occupied with two guests for the last eight days – Cottrell the Entomologist, has been on a locust preparation campaign in my area. I could have wished to be better settled in before they arrived, but it could not be helped. Apparently, there are signs of activity up in French country, and it may mean an invasion again as in 1930, I sincerely hope not (the invasion of locusts this time, not Vichyists).

Actually, I am at the Tono Farm Centre, in my little rest house on the hill in the bush, about five miles south of Navrongo. We have about 140 acres of land here, and this farm serves the Navrongo district for training farmers and farmer's bullocks in mixed farming. A cow was badly mauled by a lion a few weeks ago here; it jumped a very high wall.

I am grateful to find my policy has been followed, and made good progress during my absence. This has been an exceptionally dry year, and apparently I have missed some of the most unpleasant weather experienced for a long time. It is good to be back in the North; although this separates one from one's family, it is a hopeless sort of existence. But I feel there is a better time coming. It is a long time since I felt so happy and confident about the future, from which you will gather I am getting the mental wavelength in better tune again.

It is not surprising that after 12 years spent up here in fairly intensive work, I should find many of my interests lie here. There is big work to be done in the future. Soil conservation work on a big scale is needed. We are doing preliminary experimental work now. I hope then after the war, staff and funds will be provided in an adequate scale, and that by then we shall know our own minds well enough to go forward with a really big scheme on a scale that the situation demands. It is dinner time…"

6th August 1944 "Dear Folks,

I have had such a Sunday as I enjoy here. Blessed Day! Fiddling with lamps – the spares are beginning to become a problem now – repairing my typewriter, odd bits of reading, a haircut, a bottle of beer before lunch; small sleep after lunch. Ride over the hill to the school and back by the farm – old Humphrey is as saucy as ever but definitely ageing a bit now. Riding is a grand exercise and an excellent tonic, quite apart from the fact that it's by far the best way to see the countryside.

Alas, now I am writing under difficulties – the lamp is standing in a bowl of water and it has already trapped hundreds of insects of varying varieties; but many come on to buzz around my pen, down my neck and up my trouser leg. Fortunately, they don't bite or sting; just infuriate…"

24th August 1944 Manga "Dear Mother,

On a lovely dull, cool morning I am sitting waiting for a lorry to come to pick up my loads, servants and self, to take us to the White Volta River some 12 miles away.

I left Zuarungu last Monday week on foot, because the rivers are too high at this time of year to make it safe to swim a horse across. There are the two rivers, the Red and White Volta to cross between Zuarungu and Bawku. There is a shuttle service of lorries, which with luck and a lot of patience on the day appointed, one can get through in the day. But I came well south of the motor road, inspecting our activities and the countryside generally en route.

It was a wet trek but we were lucky with the rain, being in a rest house by the time the heavy stuff came on. On the second day, at Binaba, three and a half inches fell on Manga, and two and a half recorded at Zuarungu. This is far too much in one day to be useful, and considerable damage is caused to anti- erosion works.

Always I find trekking a tonic, and this trek has been no exception. But I know the country so well and I have so many things to do, that I really cannot afford the time to walk through the country at 15 miles a day! I have enjoyed Manga. The trees have grown up a lot, and the grass and whole area has a most settled air. The farm, too, which unfolds below the bungalow, is in better heart and working order. The crops look particularly healthy and uniform this year, partly because the other crops around are somewhat indifferent! In our unfarmable season and our methods of manuring and contour ploughing show up to best advantage.

Heard yesterday, when I cycled the three miles into Bawku, that Paris has been liberated. The chaps must have got a move on since I last heard the news in Zuarungu two days ago. We have not managed yet to eliminate the flying bomb, I am sorry to note.

If the shuttle service is working today I am going to send my boys and loads across the second river to Nangodi. Nice rest house there, now that the mine has closed down. Government took over the mine bungalow. I have a job of work to do in Zebilla, half way between the rivers. The chief of Zebilla is a mixed farmer under our scheme, and at Teski, a small village a few miles north of Zebilla, we have a newly established demonstration garden growing tomatoes, onions, spinach and beans with the odd fruit trees – guavas, mangoes, pawpaws and a dam and some anti-erosion works. I propose to leave the lorry at Zebilla, do my job of inspection at Teski, and cycle on to Nangodi, by which time my boys should have established themselves. It is a nice cool day and the exercise will do me good.

This will probably be the last year that Bawku will be cut off from the rest of the country by road; the Governor has ordered the Public Works Department to take over the road and to put in permanent bridges over the rivers. I have seen some changes in my 15 years up here. When I went up to Gambaga in 1930, to chase the wily locust, the road went only 20 miles north of Tamale in the rains. If a country is to develop it must have communications in all weathers.

I hear the distant rumble of a lorry, so must stop. The lorry is late, but these people are never very quick in the cold damp mornings. Also 'Ramadan' has started with the new moon, and many of these people play at Mohamadism and fast…"

27th August 1944 "Mother Dear,

Well, I arrived back yesterday from my trek in good order and it is pleasant to return to one's house, especially when some mail awaits. The air letter certainly gets through quickly and is a great improvement; far more satisfactory than the air graph for which I never can rely. Tomorrow, I'm off on my travels again, to Navrongo, in the other direction this time. Three nights at our farm centre at Tono, and collect the monthly pay from the treasury. Then I hope to sit in Zuarungu for a week, before going down to Gambaga district for a few days. It is far too big an area for one man to trek; a self- reliant staff of four could make some impression on the place, and maybe, after the war such a staff will materialise…"

7th September 1944 Zuarungu "Dear Mother,

Life in Zuarungu goes on much as usual. I have now settled down again. It takes two months to do so as a rule. We have been having lovely rains this last month, and the farm presents a pleasant picture of good health. The grass land in particular is thickening up remarkably, I thought as I rode around the farm this evening on the faithful Humphrey. When I think what a bare stony area it was when we started it just about 10 years ago, and look at the thick cover of vegetation today; it gives me a very satisfactory feeling deep down, which I find it hard to describe.

I wish I could report as much progress in the districts as a whole. Progress has been made, but the pastures here have still to carry too much stock, and there is no irrigation of grazing. They suffer badly during the prolonged dry season. The days seem to pass very rapidly, and I never have time enough for all I want to do.

We have had a number of visitors this last week. Major and Mrs Butler have been up from Tamale. He is acting as Chief Commissioner during Colonel Gibbs's absence on leave. He will be a good thing for the North. A new government station is to be centred on Bolgatanga (185), four miles away. A new District Commissioner's house is to be put up, a hospital, police lines, offices and all the regional Government. We will then take over the whole of Zuarungu as an agricultural station. The new site is a fine one, but I wouldn't leave Zuarungu where we know the soil and have planted trees and things. Anyway, I doubt I shall be up here when this all happens. I am hoping to get a telephone – not an extension – for use in my office here, to save our going four miles down to Bolga to phone. It will be a great boon. We are becoming quite up to date…"

10th September 1944 "Dear Mother,

The Acting Chief Commissioner suggests that I should take over the duties of District Commissioner, Mamprusi, which is a bit of a tall order in addition to my other duties, so I'm not biting immediately. If they are stuck for people I'll take it with pleasure, for it will bear out ideas which I have long held, that administrative officers in the Protectorate should have agricultural qualifications.

No bowler hats or trams to catch here. I have certainly avoided these, even if I have gone to the other extreme. But I can still remember two buckets of water being carried on yokes from a well or underground cistern at East Lound (186), and the pit latrine system, and oil lamps. Well, my water is carried from a well a quarter of a mile away. How the compound boy must bless the crazy white man who put his house on top of a hill, where there is no water! The sanitation is the bush system, and we are still lighting the pressure lamp, so perhaps I haven't strayed too far. I must say I find this quiet life very full and satisfying. I miss my family of course. But then they were never part of this life originally, and therein it is probably harder for Marjorie than for me.

I have still got the faithful Abugari as cook. He is quite remarkable…"

24th September 1944 at Gambaga "Mother dear,

Some time ago the Chief Commissioner asked me if I would like to act as the District Commissioner for Mamprusi, which comprises the Gambaga, Zuarungu and Kusasi (Bawku) districts. Well, I thought about it, and decided to take on the job – in addition to my own duties of course. An Acting Allowance of £10 a month is a consideration. Although I must confess I did not know about it until after I had accepted the job.

I am getting stale in Zuarungu; there is so much to do which cannot be tackled without more staff, funds and equipment, which we cannot get before the war is over. So I have a mind to give more time to South Mamprusi, which I last trekked thoroughly when chasing locusts in 1930. It will be no ill store to see the inside of another's job, and Gambaga is a quiet station. I shall divide my time between Zuarungu and Gambaga, coming over here for 10 days or so each month.

Another big point about taking on more work is that it will help to keep my mind off separation from the family. This loneliness grows more and more irksome.

A file of all pending letters that needed action or reply when a new officer took over a station was handed over by Anderson, the acting Assistant DC Bawku after my arrival on Friday. One of the first letters on the file was from the Paramount Chief, the Nayiri, to say a lion was caught in a trap and could he have some Government police to help his gunmen to dispatch it. So after an early cup of tea on Saturday Anderson, who was showing me how to take over the station, set off, with myself, three policemen and two of the Na's hunters. After four miles we came across the enormous footprints and the trampled grass where the brute had rested – an enormous shape. Apparently, it was carrying the trap around. Well, we followed it into the bush and the grass grew longer and was around our waists, when suddenly a growling told us our friend was about.

Then we got a glimpse of him about 60 yards away, the only sight we had, he was limping but going strong. Well, we followed him up into the grass, like walking partridges, he was growling and grunting and we expected a charge at any moment. Eventually, he led us to a place where the grass was very thick, and rocks were piled one on top of another. From this place we couldn't budge him, stones and sticks were thrown in, but he just growled. Then one of the Na's hunters went round with two police against our instructions, right up to the patch were the lion was lying. There was a noise like a train going through a station and he charged. Anderson and I were about 30 yards away, but could see nothing of what was going on except the violent movement of the grass, and men's heads and arms milling about. But the lion kept still to cover. He knocked down the Na's hunter who suffered many minor wounds, but fortunately was not killed.

Then we called everyone off before worse befell. It was no good in this thick grass as one couldn't shoot anything one could not see. When it was all over one of the policemen looked at my rifle – a police rifle borrowed for the occasion – and said 'that one is no good, it misfires'. I broke out into a cold sweat and thought we were well out of it all. It was a great pity we did not get the lion though. I don't think it will go far anyway however. We got back to breakfast at 1 pm and had done no payment for the day. A few days later we learnt that the lion had been found dead from exhaustion,"

There was certainly no conflict in the two jobs, Administration and Agriculture. I found I had to be careful when thinking aloud as an Agriculturist, that I was not taken too literally by Chiefs and headmen who usually hastened to carry out what they thought were the wishes of the District Commissioner; direct rule rather than extension teaching. The sighting of a Native Administration Dispensary with a doctor, the local Chief and the Health Councillor; or the placing of school leavers from Standard VI in the Government school at Tamale, or the White Fathers Mission at Navrongo; and the timing and sighting of Rinderpest Inoculation Camps for the local cattle with the Veterinary Officer were all matters which came naturally.

There were road gangs under foremen trained by the Public Works Department who carried out continuous repair and maintenance work on the two main gravel roads which ran through the district. The Treasury work was minimal in Gambaga, and I had two good treasury clerks who kept me on the right lines. My only complaint was when the Government mail lorry arrived, with specie on board, in the middle of the night and I was called to open the vault which was located in the prison.

4th October 1944 Zuarungu "Dear Mother,

Here is a go at an air letter from our end. I am mighty busy these days having become an Acting District Commissioner – magistrate, coroner and sub- accountant overnight – without any knowledge of the details of my job, in addition to my normal multifarious duties. But it is all experience and I would not have it otherwise. Nothing like plenty of work to do in these strange days!

We have started haymaking, and Zuarungu is full of nostalgic scents and scenes. Couldn't resist the temptation to get up and stack this morning, much to the amusement of the labourers. But in half an hour I was in a muck sweat, and sneezing to beat the band. There is something almost sacred in farm work. On these occasions I am inclined to wonder if I shouldn't have stuck to more temperate climes, where the white man can work with his hands with dignity.

My work makes some progress. We are held up now with shortage of staff and equipment; that is the reason I am able to take on the extra duties. By so doing I hope to be able to review agriculture in Mamprusi more distinctly and to plan a policy for the time when staff and equipment can be obtained.

Tomorrow I trek up to Bawku. I return on Tuesday, and on Friday go down to Tamale for a long weekend. Colonel Gibbs is back in Tamale as Chief Commissioner I hear. From Tamale I go to Gambaga and return to Zuarungu at the end of the month. I have now got three residences – Zuarungu, Manga and Gambaga, and a rest house at Tono. I like living in all of them.

Now 'tis time to move further away from the lamp, for the insects are beginning to fly and there seems to be a spate of particularly vicious cantharides blister beetles, capable of raising the most unpleasant blisters if one accidentally crushes one against the skin…"

23rd October 1944 Gambaga "Dear Mother,

I am as busy as ever in my dual job, and must say I am enjoying it. Much time is spent poring over the Criminal Code and Code of Criminal Procedures. So far I have only appeared on the bench to adjourn a couple of cases. I pray there are no murderers, or an inquest to be dealt with. It's all experience. I have to pay out soldiers' families, pensions, and so on in addition to all the other Government payments here. The treasury work is new to me from the inside. Fortunately, I have two office cashiers. Agriculture is not being neglected, because I have long wanted to start an agricultural survey and agricultural register for this area, and this is my opportunity.

Gambaga is a pleasant place. My love of the NT remains. As far as bush life is concerned, it has changed a bit since I first knew it, and is much more developed now, with extra roads, and a post office. Having become a senior station it is very well equipped and well staffed, which makes just all the difference…"

Court work was not too onerous; I had been sworn in as a Magistrate when last in Tamale. I had one or two traffic offences in Walwale, with the accused invariably pleaded guilty by telegraph. For other cases I would sit on the Bench with the Criminal Code and Book of Criminal Procedure in front of me. Serious cases were remanded to Tamale for the Assize Court. The Chiefs' Courts tried all cases of native law and custom, but referred criminal cases to me. On one occasion, we were driven out of the court by a swarm of bees and had to hold court on the veranda, only the accused had the presence of mind to warn me of the imminence of the roof truss by tapping his head. We had also to compete with the braying of a donkey, which was hobbled by the front legs to graze outside the courthouse.

When on tour in the district there were Native Administration court records and treasury books to be examined, and family allotments to pay made for men in the Forces. This was not a straightforward procedure, because many men registered a completely different name to the one by which they had been known at home. It was quite a game tracing everybody, requiring the presence of the Chief and the whole village. Even so it was doubtful whether all allocations were correctly made.

12th February 1945 "Mother dear,
We are having a fine spot of cold harmattan weather this year, which is very pleasant. I'm hoping the hot season before the rains will be a short one – it varies very much from year to year. Farming matters do not seem to have suffered from my neglect in this area these last few months.

I have a fairly full programme before me now; until I am due to take over in Tamale. I'm glad I came back to the north for this tour. I feel my leave- taking will be gentle. I have contributed as much as I can to affairs here, and feel I can cheerfully hand over the outfit to another now."

18th February 1945 Zuarungu "Dear Folks,
I have been doing a lot of trekking these days – a District Officer must move around everywhere, and in the dry season there is not so much going on at the farm. We are still continuing with our bucket- rearing of calves

here, and milking out the cows. Our best so far is a gallon of milk per day, and 200 gallons in a lactation period. But of course our cows weigh only about 4 cwts.

Building repairs are the big works of the dry season, and the buildings we put up with sun- dried bricks and thatch, though cheap at first, cost and require a good deal of maintenance. The big event of the month has been the re- opening, at long last, of the Government treasury here, which was closed down soon after I came in 1932. And yesterday the chaps finished putting in the telephone line to my office. They have put up a line from the P.O. in Bolga four miles away. It will be a great boon; it has taken an awful long time to get these quite ordinary facilities installed. Odd they should arrive just before I leave.

I doubt if I shall be stationed again in Zuarungu. My leave- taking will be gentle. I have become somewhat stale, and feel that I have done all I can contribute to the Fra-fras. It is somebody else's turn now to carry things on further, and I am happy in the knowledge that Roland Smith takes over from me, for he will still farm where I leave off, and not from the beginning again.

Tomorrow, I am off again to Navrongo this time. On Tuesday and Wednesday, I have a date at Chana Rest House with the Forestry Officer. He wants to declare a watershed there as a Forestry Reserve, an idea with which I am all in favour. We are going to get together, and discuss the boundaries, farming requirements in mind. We don't work together, in spite of all the talk of co- operation between departments. It is essential to any land planning projects, and these are so much in the air…"

25th February 1945 Zuarungu "Mother dear,

I have had an interesting week spent mostly at Chana with the Forestry Officer and DC Navrongo, going into the demarcation of a forest reserve around some neighbouring hills. We rode horses up the valleys, and then when it got too rocky climbed on foot. It was very pleasant indeed. Amazing how the Chana people will walk three or four miles to plant up little pockets of soil in amongst the rocks, yet find it too much to walk a hundred yards to cut bedding grass to make manure. Part of the trouble is that by wasteful farming methods they have worked out the lower levels, and are now forced up into the watersheds, with disastrous results for the water supplies. Officers of different departments don't get together often enough.

My colleague Roland Smith comes down from Manga today to spend a few days with me. Tomorrow, we go to Bongo to discuss the question of stopping farming up in the hills there. The run off from the top has produced an enormous gully about which we are concerned . It is experimental land planning. We must get our data upon which to base our soil schemes. The people at the top don't seem to worry about this. All the schemes they seem to regard as an end in themselves. The real needs of the people are comparatively secondary.

We took the Chief of Bongo (187) with us, and enjoyed it all very much, and settled everything amicably on the spot. No doubt we have at least won the confidence of these people, in the years we have been here, even if progress has not always been what could be desired. The faithful horse Humphrey still carries me about in style. He is getting a bit over at the knees now, but can pull as hard as ever.

There is still a cool harmattan wind blowing, and two light blankets are the order of the night. It is so often stinking hot by this time. If the rains break early it will be grand. It makes such a difference to escape the worst days of March. I have a pretty extensive trekking programme to get through this month. Early in April I was due to go to Tamale to take over from Broach. Having got settled in here again now I feel rather sorry to be leaving just as the things are growing…"

The prison was a useful institution, generally with occupancy of from none to three. Removing latrine buckets from the District Commissioner's house and rest houses, and cutting firewood in the bush were main occupations for the prisoners. On one occasion when the prison firewood party was in the bush, the warder fainted, and was carried back to the station with his rifle by the prisoners, splendid esprit de corps. There was no great stigma attached to prison life, where clothing was provided and food regular. I received a shock one day when a telegram arrived from Accra asking what the financial implications would be if Diet No. 3 was substituted for Diet No. 2, as laid down in Prison regulations. Since the current diet bore no relation to either I had to send a noncommittal reply, and muse on the wonders of bureaucracy.

Being a District Commissioner I found to be pleasant with never a dull moment, rewarding but exacting, one's responsibility for all Government affairs never ending. One never knew from which direction a problem would arise. I never got to like the heavy treads of the police sergeant on the gravel, when I was resting on Saturday afternoon. I might be told there was a body in the mortuary and would there be an inquest? Then the doctor would have to be fetched from Navrongo 60 miles away.

There were also problems with the roads, when lorries went

through the bush stick culverts. Army recruiting lorries were often the culprits, because they were the heaviest vehicles on the 30 miles of road between Walwale and Gambaga. On one occasion, a highly indignant American missionary from Walwale, riding a motor cycle with his wife on the pillion, came to grief on a broken culvert. He complained bitterly in my office that his wife had broken her "braisier".

I had a strong inkling that after a spell of 13 years my sojourn in Zuarungu was coming to an end. Much had been achieved in that time. I decided to take the opportunity to visit Lawra, Wa and Bole before going to Tamale because very soon the rains would cut off the west side. Johnny Hinds was established as the Agricultural Officer at Babile in Lawra district, with Robert Talbot the District Commissioner. I enjoyed attending a Native Administration estimates meeting in Lawra, and hearing the Lawrana strongly defending some action of Johnny Hinds.

I moved on to enjoy the hospitality of Cecil Amory, District Commissioner at Wa. It was then I heard that Marjorie and Sylvia were on their way up from South Africa by air. They were expected to arrive in Accra on Wednesday 18th April. The family duly arrived, but had experienced six worrying days in Leopoldville and three in Lagos waiting for planes. I heard that my old friend, Kibi Sir William Jones at Achimota, was looking for a home for his two- year old bull terrier bitch "Jill", a telephone call and Jill was delivered to the train. So I proceeded north with a wife, a daughter and a new dog, which became an immediate favourite particularly with Sylvia.

It was a felicitous homecoming with all the family to Tamale, with its many amenities. The Medical Department said they would accept no responsibility for Sylvia, then nearly three years old. But the local doctor Benny Goodman told us not to hesitate to send straight for him if we felt the need to do so. The only other white children in the Northern Territories were Michael Syme, a little older than Sylvia, whose father Jim Syme was the District Commissioner at Salaga, some 80 miles south of Tamale; also the one- year old son of Hurtado, a Spanish trader in the town, and two children belonging to American Missionaries.

We were fortunate living out on the Experiment Station, where we had a very healthy life. There were many things to interest a child on the farm. It was my custom to ride round at 6.30 every morning on the faithful Humphrey, with Sylvia in front on the arch of the saddle, and the bull terrier Jill trotting behind. Sylvia also had her own donkey to ride.

29th April 1945 from MARJORIE "Dear Mother,

I still have to pinch myself occasionally to convince myself that it is real and we are here. It seems very strange to hear the patter of childish feet in these houses, where no children have ever been. The Symes will soon be in Tamale with their little boy of five. There are two missionary children, but that is all…"

14th May 1945 "Mother dear,

For me, the last month has been more of a whirlwind than life usually is out here these days, with an ambitious programme and shortage of staff.

I was in the midst of a week of toil when I sent my last scrappy air letter. My labours were brought to a speedy and hurried close by an urgent personal telegram from the Director. So Michael Akenhead, with the sheets of my six year plan still wet from the typewriter, leaving yesterday at 2 pm. and the final appendix still in draft; he dashed down the road with it to Kumasi. Unfortunately, he broke two springs on the way and arrived two hours after the Accra train had left. Bad luck after a good effort. However, I was able to phone through the main figures and the scheme will go down on tomorrow's train with a personal messenger.

So it goes in this crazy time, six years planning in three weeks, spending £486,000 on paper. It is a pretty cast- iron case, and should qualify for a grant from the Colonial Development Fund if the Colonial Government can't afford it. Well, we'll see. It was a great relief getting it finished and I am quite satisfied with the result. My next problem is to produce a plan for the Agricultural Resettlement of Demobilised Soldiers, in time for the Chief Commissioner to take to Accra on Thursday.

But it is a pleasant change to be in Tamale, with all its amenities and wider scope – electric fan, telephones, running water and all manner of facilities. The farm is doing well. I was writing the monthly report for April this afternoon. It is quite impressive in a small way. 60 lbs worth of vegetables from the garden. 500 gallons of groundnut oil, 290 dozen new laid eggs, farm fat sheep, three farm pork pigs, and 30 weaners for finishing in the south. 60 lbs of butter, seven pints of cream, and 160 chickens – mainly for local army supply. Also 20 bales of 100 lbs each of kapok for the Ministry of Supply. Thank goodness that the kapok will not be needed for the lifebelts in the Atlantic and Mediterranean now. The way things are going it looks as if the Pacific theatre too may soon be cleaned up…"

30th May 1945 "Dear Mother,

The vast plans have received the general blessing of His Excellency the Governor, so future developments will be interesting. The whole Colonial Agricultural service is very understaffed, as who they are going to get to put these plans into operation heaven knows. The Colonial Service has been very meanly treated in my opinion, and it will only be a falling- off in recruitment, which may bring them to their senses. If they looked after their staff properly and told them the truth on appointment, there would be less invaliding and dissatisfaction…"

2nd June 1945 from MARJORIE "Dear Mother,

The birthday season has passed. Your cable arrived midway between the two, and was duly appreciated. I managed to arrange a small party for Sylvia – we had a little boy of nearly five staying with us for a week, then there was a small infant belonging to some Spaniards, and two little Americans. The latter brought a little packet of chewing gum tied up with yellow satin ribbon as their gift! Those were all the children I could muster, but several adults came as well…"

There were cattle, sheep, pigs and poultry kept commercially for army and civilian supply as well as the large vegetable garden behind the Tamale dam. My predecessor Jimmy Broach, now on leave, was an excellent practical farmer and had left me a productive going concern. Of course, we were able to achieve this direct production with excellent African staff, they were splendid people, hard working and trustworthy, with a sense of humour.

19th June 1945 "Dear Mother,

Last week, we were on trek around on the west side of the NTs. Firstly, Lawra and Wa, coming round the North through Tumu to Zuarungu and back. I was not anxious to take all the family. We motored over 600 miles, but they insisted on coming. Fortunately, the car went well or it would have been awkward.

The only snag about going away is the pile of correspondence and routine stuff to be dealt with upon one's return. This week has been no exception. The staff question is very difficult now. Michael Akenhead has gone into hospital with amoebic dysentery, and he is due to go on leave early in August. No relief is available so Roland Smith, in Zuarungu, and I here will be left to hold the fort. However, I would not have it otherwise if it meant being in Tamale for this pleasant interlude.

I managed to get some of this wonderful DDT stuff. (188) It is particularly potent against the malarial mosquito. We have just given the house a good spraying. The effect is supposed to last for a month or two.

Another new toy is an electric fence which works very well. I am sending one up to Zuarungu, to get my own back on the local goat..."

3rd July 1945 "Dear Mother,

The cost of living here is wicked, and the good kind Government does nothing to help the civil servant. They paid only £57 out of £132, being Marjorie's fare up from the Cape.

The wonderful stuff DDT is still keeping the bungalow free from mosquitoes. It will revolutionise life in some places if it goes like this..."

12th July 1945 "Dear Mother,

Michael Akenhead is just going on leave after a long tour and Roland Smith and I are left, with a fantastic programme of work on hand. I have 18 students to track on top of everything else. An ex-soldiers resettlement scheme which I put up subject to staff being available, has been warmly approved in toto in Accra, but no mention of staff. I have now been told to recast my six year plan on a 10 year basis, and to make it more modest as regards district staff. But it's all good fun so long as it doesn't get you down. The Colonial Service has gone crazy. They have reduced our salaries by allowing prices to rise by anything up to 500%. They will not raise a hand to assist people to be joined by their wives. The salaries offered now are not attracting new recruits – which pleases me in a way – bungalows are in a filthy state. At the same time we are all having extra planning pushed onto us, in which we have no confidence. The native policy, too, is disquieting. We are pushing it much too fast and all development is taking place in the towns..."

The war in Europe was ended. Development plans for a 10 year period were urgently required to be financed from the new Colonial Development & Welfare Fund, recently established by Government. Soldier Settlement Schemes were dreamed up to deal positively with the problem of reabsorption after demobilisation.

Another important change was coming too, as up to now life was blissfully free from politics. British officials exercised a benign form of direct rule through the hereditary hierarchy of Chiefs and headmen, on the lines of Lord Lugard's "Dual Mandate". It was a system which worked economically and well. Increasingly, the Chiefs and their councils were given more power and responsibility, through the Native Authorities, which in effect were the local government authorities.

The thinking was to learn central government through local government. It was true that the system was not democratic by

Western standards, but it had to be remembered that Chiefs ruled by the consent of their people, sudden death being by no means unknown for an unpopular Chief. People paid a tribute to their Chief in the form of produce and labour. The Chiefs, in their turn, offered hospitality to strangers, and helped out in famine times with food and seed. This close- knit structure of family, clan and tribe made for stability and continuity. Europeans could in no circumstances own land. A policy of progressive Africanisation was always in operation as educational processes had their effect.

For many decades the Gold Coast had the benefit of local professional men, mainly lawyers and doctors holding normal English degrees. During the thirties an increasing number of local Agriculturists, Veterinarians, Educationists and Administrators were obtaining their degrees in British Universities. Then they returned to take their place on the senior staffs of the government service. With the return of troops from overseas service, and the infiltration of leftist ideas from Britain, there was a growing tension in the towns in the south of the Gold Coast and to a lesser extent in Ashanti.

There was restlessness with the old traditional order of things, and a cry for democracy with "one man one vote" suddenly. Political forces were at work for which the Government was quite unprepared. In the Northern Territories we were comparatively unaffected by these changes, but although we were unaware of it at the time, benign rule through the District Commissioners and Chiefs was coming to an end. Stability was replaced with uncertainty. (189)

22nd July 1945 "Dear Mother,

It is a lovely dizzily cold day here, such as one occasionally gets in the rains to remind us of home. We have been extremely fortunate. The season has had good rains. Crops are growing well, and stock is in good form.

We had Mrs Kerr staying with us this week. Allan Kerr was my neighbour in Zuarungu for many years. So we have spread ourselves a bit – been to the pictures twice – pleasant in the open. One puts up one's own camp chair, and sits on the roof of the old pavilion. It's really a show for the military of course. The snag is we get no advanced information of what is coming on, as this means watching a lot of trash; but even so it is a pleasant change, and first rate recreation. Tennis, too, is gaining popularity again in Tamale, and we play two evenings a week. I am as busy as ever, having started regular lectures for my 18 students. Soldiers resettlement schemes too, occupy much of my time, but it is cool and easy to be usefully busy in such conditions. No doubt having one's family around gives one the feeling of stability, with a better balanced outlook on life.

Michael Akenhead has gone south to catch a plane on leave. There is only Roland Smith and I left at the moment in the NTs. Quite fantastic. The staff position is becoming serious. I see in a recent 'Farmer and Stockbreeder' they are advertising the Colonial Agricultural Service, but the starting salary is the same as jobs advertised at home for people with similar qualifications. I would not advise anyone to come into the service now with the present cost of living and the mean policy being pursued by the Government towards its servants"

2nd August 1945 "Dear Mother,

Perhaps our new Government may give us a more enlightened Colonial Policy. It couldn't do worse. Cut out all this window dressing and pretence – a cloak to big business projects – and fulfil our trumpeted obligations to these people. At present we are raising them up only so they will fall. What a sweeping victory for Labour.

Good luck to them – they must have had the backing of the Home Guard, and that's good enough for me. I am getting cruel joy out of the reactions of some of the men dyed- in- the- wool Colonial Officials. One might think the end of the world had come. The new Government has some brains and heart in it and good luck to 'em…"

In the middle of August 1945, my Director suggested that I should come down and meet him for discussions; he believed in the value of personal contact.

15th August 1945 Tamale "Dear Mother,

I have been called down to Ejura – 170 miles south, to meet my Director tomorrow morning. I duly set off in the car this morning. After going 10 miles I noticed the back of the car sinking, and decided that discretion was the better part of valour. I was afraid that the back spring was about to break. So here I am awaiting a kit car from the Government Transport Department. It's a bit hard on VJ day to have to paddle round the country on duty bent.

I have been busy as usual especially getting out 1946–47 estimates; it makes life seem very swift- moving, thinking so far ahead.

Ejura Rest House 7.45 pm VJ DAY "Dear Mother,

Well, here I'm in this old rest house again, on top of the scarp in which I used to live in 1933. The lorry came down very well. The road was not too good and we ran through three storms. The rain is pouring down now.

We got in just in time to get out the lamps. Once again I'm comfortably settled in my rhorky chair with my book box beside me, the pressure lamp going and in the next room a camp bed and canvas bath. This is rather a palatial rest house, since in the early days it was the residence of the DC A good thing it is watertight – at least it appears to be. There are due to be the usual celebrations in Tamale tonight and on Friday which I am missing and not minding either…"

23rd August 1945 "Dear Mother,

Well, I got back safe and sound, having enjoyed a morning with the Director. What a change to meet one's Director and feel stimulated instead of frustrated. No doubt personal contacts are good. He told me, amongst other things, that I was not to bank upon Jimmy Broach returning to Tamale after leave. Well, if it means I stay in the North it suits me. It is the only part of the Gold Coast which provides a life in which I can be fully interested.

My word it is raining; rained all night and it is still coming down in bucketsfull. Too much this year; often more harmful on crop yields than not enough; but it's cool and Sylvia keeps remarkably well. Indeed, I enjoyed seeing her bobbing about this morning, playing with the dog. It's only malaria that scares me, but provided we keep her out of reach of mosquitoes, and here DDT comes into the picture.

I rush around as usual, trying to reconcile day- to- day details with the Ten Year Plan. On Monday, I have to accompany the Acting Governor up to Zuarungu, Bongo and Navrongo, returning Wednesday. He is coming up to see farms and farming, and to settle the Veterinary and Agricultural question. I was particularly pleased to be able to see Urquhart before this visit, to know whether he wanted me to put on the 'whitewash', or to take it off. I am pleased to say he said no 'whitewash'.

Johnny Hinds has arrived back from leave. He came up the road yesterday from Kumasi and will go round to Lawra. Roland Smith is bringing his wife up today or tomorrow, whenever he gets his car fixed. All cars are getting old and groggy these days…"

The following week, I went north to Zuarungu with a party consisting of the Acting Governor, Mr Gurney (later Sir Henry Gurney, Governor of Malaya), my old friend from the R.W.A.F.F. Rex Ames, now a policeman ADC and Guthrie Hall, Assistant Chief Commissioner. The purpose of the visit was to see something of the countryside and the work of the Department of Agriculture in particular.

We showed His Excellency the tobacco beds and vegetable gardens at Winkogo and Bolgatanga, and crossed the Bolga River by drift, since the wooden bridge was broken. Then went on to

Zuarungu to see the farm centre and have lunch. Fortunately, we sent the cars back to Bolgatanga in case the storm, which looked imminent, washed the drift away. The storm came and sure enough the drift was under water but not too deep to wade. Mr Gurney did not hesitate; he removed his shoes and socks, rolled up the legs of his immaculate Palm Beach suit and the gubernatorial party walked across the drift in strict order of seniority with Rex Ames and myself bringing up the rear! The Acting Governor behaved as if he did that kind of thing every day of the week. Very thoughtfully upon his return Mr Gurney sent up a large supply of watering cans, for distribution to vegetable growers.

19th September 1945 Tamale "Dear Mother,

We have had a bumper crop of mosquitoes this rainy season. I have been chained to my office for some time now until I cut a dash. Colonel Steemson, my opposite number in Ashanti, came up for a look around last week and I seized the opportunity to go east to Yenli and north to Zuarungu and Navrongo. Everything was looking very flourishing, and the change did me good. I feel a bit tired these days. There is so much to do, but it is rather unsatisfactory not knowing what my fate will be when Jimmy Broach returns. His leave expires in three days, so he will be back quite soon.

I expect Urquhart up next month, so should soon know my fate. I should be very sorry to leave my beloved NT and would certainly not like to be moved this tour. However, we must be patient and wait and see…"

1st October 1945 from MARJORIE "Dear Mother,

Now there is a lull and Charles has been ordered a few days rest by the doctor. He's just got to try and forget work and read and sleep. He hasn't been terribly fit for the last few weeks. He has had a very hard tour, and the last two months he has been doing about three men's work. A week's rest will do him the world of good. He takes all his work to heart so much as you know. Later on we hope to take some local leave and go to the sea. These long tours out here, that we have had to do during the war, are no good. So many men are feeling the strain…"

9th November 1945 Accra "Dear Mother,

Well, here we are, enjoying two weeks of the amenities of Accra combining local leave with a very small amount of work. Mainly useful contacts and discussions…"

19th December 1945 Zuarungu "Dear Mother,

Roland Smith has had to go to Navrongo to see the doctor about a tooth, and may have to toil down to Kumasi 350 miles away to see a dentist if the doctor can't fix it.

I have been on trek for just over a week, leaving my womenfolk behind in making this trek. I went across from Tamale to Wa on Monday, where Johnny Hinds met me. Up to the farm at Babile, near Lawra, where Johnny lives. Tumu on Thursday, Tono Farm near Navrongo on Friday, Manga on Sunday. Down here this morning and back to Tamale tomorrow. A fairly comprehensive and very satisfactory tour. We have greatly changed the countryside for the better in the last 10 years – hundreds of square miles of grass are now protected, when before they were burnt off as soon as the dry season came in November.

It is very dry up here and wonderfully cold at nights. I had a lovely ride around my old haunts this evening on Roland's pony 'Patience' – an Accra Race Horse and a lovely ride.

So I am feeling pretty 'jake' tonight, I don't mind telling you, with 38 little jobs in my note book to receive attention when I get back. But it is fine to feel really fit again – that local leave was very much worthwhile…"

The old experiment station, on the north road two miles out of Tamale, was far too small for future mixed farming needs. So during the late rains and early dry season in 1945 I endeavoured to find a more suitable site for a central agricultural experiment station. I enjoyed horse or bicycle rides in the early morning, often accompanied by Guthrie Hall, the Assistant Chief Commissioner NTs. We were looking for a station with soils typical of the Voltaian sandstone region, preferably served by a road, within 10 miles of Tamale. We needed 3000 acres at least, free of villages and with topography which allowed at least one dam site. We found just such a site at Nyankpala, at mile 12 on the Yapei road. It had the advantage that the Nyankpalane was known to us as Mr Damba. He was the urbane Head Overseer of Tamale station when I first arrived 15 years before, since retired and appointed as Chief. He was pleased to have us as neighbours and was a great help.

Christmas 1945 was a happy time with a fancy dress dance at the club, the traditional dinner at the Residency and a Christmas Eve Carol Party on the Agricultural Ridge around a huge log fire on the lawn. Having one's family with one made all the difference, so many were sadly separated, part of the price paid for service in West Africa.

Immediately after Christmas we had a conference of senior officers of the Department to discuss policy and plans for the future. Present were Michael Akenhead, Roland Smith from North Mamprusi, Johnny Hinds from Lawra and Chris Dady, a locally appointed agricultural officer who had just completed the Natural Science Tripos at Cambridge. We were joined for part of our discussion by Guthrie Hall, Assistant Chief Commissioner, Allan Kerr, DC Tamale and Kenneth Dickens from the Department of Education. We enjoyed a far- ranging discussion of the whole field of agricultural development for the Northern Territories.

To my joy polo was restarted in Tamale in January 1946, after a lapse of six war years. We played four chukkas an evening on Monday, Wednesday and Friday on the military parade ground, and it was like old times.

The staff position was greatly eased by the posting of Chris Dady to the North. Life for the officer in charge was not without its special problems; the Agricultural Officer Dagomba was building a rest house on the new farm centre at Yendi, which incorporated a cellar to serve as a storeroom but which attracted frogs and snakes. At the same time, the Agricultural Officer, Lawra, was building a house at Babile on split levels. There was a need for a common government policy giving a combined approach to development, co-ordinated by the administration. District and Provincial Development Teams were coming into being, to resolve this kind of problem.

In March 1946, Broach returned from leave and was posted temporarily to Tamale to relieve me. Shipping was, of course, very uncertain in those days and there were medical priorities to get away first. I was told to report to Sekondi for special duty with the Chief Agricultural Officer, Western Province, Colonel B.T. Steemson, recently released from commanding a battalion of the Gold Coast Regiment. My special duty was to try to locate a suitable site for a headquarter station to serve the Western Province, various possible sites having been suggested.

We were ultimately offered passage on the P&0 ship "Ranchi" sailing from Takoradi on 15th May bound for Tilbury. The voyage home was enjoyable, although the "Ranchi" was still a troopship and accommodation was segregated. Sylvia had her fourth birthday on board.

Sir Harold Tempany, the Agricultural Adviser to the Secretary of State for the Colonies, called me to see him in London. He told me he proposed to second me for special duty at the Imperial College of Tropical Agriculture in Trinidad, for three years. A new course in Agricultural Extension for postgraduates coming into the Colonial Agricultural Service and for diploma students destined for appointments in the West Indies. Before taking up the appointment as Senior Lecturer it would be necessary to spend a year visiting East Africa, Britain, Canada, the USA and Puerto Rico to make a comparative study of extension work. I was to be away from the Gold Coast for four years on secondment.

Chapter 10 - 9th Tour 1946–47

Nyankpala, Empire Parliamentary Delegation, and so Goodbye.

Arrangements were not complete for my new adventure at the end of my leave in November 1946. I returned to the Gold Coast alone, aboard the M.V. "Tamele", one of Elder Dempster's new passenger cum cargo ships – she was called "Tamele", and not "Tamale", as a result of mistaken information from one of Elder Dempster's captains. With the exception of the "Aba", used as a hospital ship in the Mediterranean, the whole of Elder Dempster's fleet – Abosso, Accra, Apapa, Abinsa, Appam and Adda – were lost in the war. I landed at Takoradi to be met by my faithful cook Abugari. I was sent north again to hand over my duties. I could not have wished it otherwise, than to be given the opportunity to ensure that such progress as we had made in the north was not lost. The Harrigan Report on salaries and conditions of service had just been published and was generally welcomed by the service.

2nd January 1947 Tamale "Dear Mother,

A large turkey seemed rather much for us over Christmas – Montague, Johnny Hinds and Eccles a new recruit down from Zuarungu. For New Year we had a great ham from the Accra Pig Farm, upon which I shall be living for the next month.

We have had a nice return of the harmattan and it is beautifully cool at nights and in the early mornings. I am greatly enjoying being back in Tamale. I am trekking about all round my old haunts – Gambaga, Zuarungu, Manga, Tono and Babile.

We have got to be back on the 15th in order to meet eight MPs who are arriving in Tamale by air on the 16th. I haven't heard who is coming up yet except Lord Llewellyn. It is a bad time of year to show anyone around the farm. But we'll try to convince them of a rural bias in the North.

I am finding plenty to do here. Broach was snatched away to act as Deputy Director at a moment's notice, and Montague is quite new up here. I have been out of touch now for nine months and there have been a few changes..."

16th January 1947 Tamale "Dear Mother,

We got back safely from trek yesterday having done 900 miles without a falter. It was good to see the old haunts again and so many 'keut' faces. Steady progress has been made, but I realise now that had I remained any longer there would be a danger of stagnation.

The offer has come through from the Colonial Office. It is a lectureship in Trinidad for three years on secondment. Lecturer in Agriculture and Extension and Advisory work – with a year's training period beforehand. Leave is given every other year for the College. It sounds pretty good to me.

We, that is Montague and I have 6 MPs coming to tea this afternoon, followed by a quick look around the farm. They are coming to us straight from the airfield. They will see quite a bit of us during their two days visit, for we are dining with them tonight and taking sherry at the Residency tomorrow. I wonder if any of them have been on a farm before. Four Labour, two Conservative and one Unionist…"

The visit started on the Tamale Agricultural Station by the Empire Parliamentary Delegation, which was touring West Africa. It consisted of Lord Llewellyn, Lieutenant L.J.Callaghan, (190) C.W. Dumpleton, Colonel F.J. Errol, E.H. Keeling, Dr Segal and F.N. Skinnard. Montague and I gave them tea and explained our policy, followed by a quick look round the farm. In the evening, we met again at the Residency with the Chief Commissioner, Guthrie Hall. The following day, I begged a lift in the Dakota, taking the party to Navrongo for the day: I was anxious to see from the air the country I knew so well on the ground. I was not disappointed, and felt that land-use planning in the future should be done by aerial survey. The control of grass burning was excellent in Navrongo and Zuarungu districts.

After a trip to Bolga Market in the morning, I enjoyed talking to Mr Callaghan in the shade of the veranda of the District Commissioner's house in the afternoon. The rest of the party went up to the School and Cathedral at the White Fathers Mission.

On our way back to Tamale in the evening it was noticed that the sun was going down on the wrong side of the plane. We had lost our way and had to return to the Nabogo River to pick up the Daboya road, which led us in to the landing strip. (191) It was almost dark when we got down, and various cars had their headlights switched on to show us the landing strip. Callaghan was singing as we landed, and Colonel Errol told him he was indulging in a selfish pastime! It was dark by the time we bumped to a halt on the dusty laterite surface.

I was able to devote special attention to the development of the new headquarter site at Nyankpala. Test pits in the valley showed a source of clay to provide a puddle core for a dam, which was the key to the water supply for the station. Lilly, Director of Water Supply Department, approved of the first dam site in the Northern Valley, and thought there was a 50:50 chance of a second dam in the Southern Valley being successful, which was encouraging. A Fordson tractor and a scraper arrived from Accra and we set to work. Polo twice a week and tennis at the club provided recreation.

4th February 1947 Tamale "Dear Mother,

How time flies. Fortunately, I am at the beginning of my tour or my wheels would be spinning, the engine racing and we'd be getting nowhere. As it is I am tolerably busy – the only way to be in this country – with so much to do and time running out.

I am writing the annual report for the Protectorate, probably for the last time. I've also got a hundred labourers working on the new central station site 10 miles out of Tamale. A new tractor is on its way up from Accra and I hope to get the chaps onto the start of a pond before I leave.

At the moment, they are cleaning the boundaries, which I am adjusting slightly. The area of the place is 4 sq miles, so it takes a bit of getting around.

It is a trifle irksome not being in charge, and not being able to initiate anything. Not that my companion Montague is not one of the very best, and a good friend of mine, but he is only Acting Assistant Director, and I am merely advising him. But I am glad to have come out to take a gentle parting from the place, to settle my affairs and get my mind attuned to colonial agriculture before venturing in East Africa…"

10th February 1947 Tamale "Dear Mother,

It is a soft sunrise, the air is warm, and there is a smell of rain about. At least we had a drop the night before last, the first for over three months. But this is nothing compared with the winter grip in which you are held at home. I am feeling a different person after two month's sojourn in West

241

Africa. We didn't regard it as a health resort in the old days.

The sun is getting up and prepared for the day. I have a date at our new central station site, Nyankpala. I have a hundred labourers working on outlining traces and digging test pits in the valley for clay. I want to get started on a pond before the rains come. We have just got a new tractor delivered, exactly as I had ordered. Actually, it came for another Department who did not want it, and we have collared it. The new area is 4 sq miles and I have to acquire the land, assess the compensation and site a training centre and other buildings before I depart…"

3rd March 1947 "Dear Mother,

I had a holiday this week at the request of the Chief Commissioner. I was invited to Zuarungu, on the occasion of the 'presentation' of the Kings Medal to the Chief of Bolgatanga. It was very pleasant up there, seeing so many familiar faces and they have been wonderful over their control of grass burning. On the way down I broke my journey at Pong Tamale, to stay the night with Jock Stewart. It was most pleasant. I am still busy at the new Experimental station site…"

14th March 1947 Kumasi "Dear Mother,

I got a signal on Tuesday last to proceed with due haste to Accra, but I still do not know just when I am leaving, nor by what route. I drove down here from Tamale yesterday after taking fond farewells with the kindly folk up there. Unfortunately, I bust a spring on the way down in the old car, but managed to get a lift on a Public Works lorry.

I left the faithful Abugari in Tamale and have Awari with me. I think Abugari intends to go back and farm in his village. He has been a grand servant. I have settled all my business here – or nearly all. It has been a bit of a scramble getting everything sorted out and finished off as you can imagine. But not nearly such a problem as it is at the end of a tour when one is tired…"

20th March 1947 "Dear Mother,

I cannot remember where I was when I last wrote. The last couple of weeks have been very messy, packing and repacking, and attending to the hundred and one things which seem to need attention when one is packing up for a long time, and sending kit in three directions. But everything is pretty well finished now, and I'm more than ready to push off. I have been told to get to Nairobi as soon as possible.

My leave- taking of the Gold Coast will be gentle and pleasant and without regrets. It does not seem so long since I arrived in Accra on my first tour and I have recaptured, in reverse, many of my early reactions. I'm ready now for the next adventure. I was sorry to part with my ancient car in Kumasi. It had done well – 83,000 miles and sold for £35 more than I gave for it 10 years ago…"

On March 31st, I left Accra on Pan Am "Constellation" Flight for Johannesburg via Leopoldville. I played polo on my last evening in Accra on ponies supplied by Jack Faux, who at one time was Assistant District Commissioner in Zuarungu and was now with the Information Department. Nearly 17 happy years in the Gold Coast came to an end, and an exciting new chapter was about to begin. I was, however, a little sad and knew things would never again be quite as I had known them.

Background stories and supplementary notes

Chapter I - 1908–31

1. *Wye Agricultural College.* Founded in 1894, and closed as an Agricultural College in 2005. After the Second World War it became part of London University, later merging with Imperial College, London with more emphasis on business skills.

2. *The Colonial Agricultural Service.* Ref. A short History of Agriculture in the British Colonies; Masefield.

3. *Imperial College of Tropical Agriculture*, Trinidad. Was opened in 1921 with the first scholars leaving in 1925. Professor C.Y. Shepherd visited the Gold Coast to advise on the Co-operative movement for the cocoa industry. In 1948, he was at the College, when Charles took up a three year secondment to teach Extension Work.

4. *Agricultural Adviser to the Secretary of State.* Ref. How Green was our Empire; Barringer P26 OSPA Research Project Vol.2, 2005.

5. *Mosquito boots.* Were knee- length boots, often of soft patent leather, worn in the evening to keep off mosquitoes and other biting insects.

6. *Chop boxes.* Were wooden boxes with interior partitions for food etc., carried on heads of porters, and fitted with hasp, staple and padlock.

7. *The Aba of the Elder Demster Line*. For a history of the Line. Ref. The Conquest of the Niger p.243, Hollett Letters from the Gold Coast, p.1 Princess Marie Louise.

8. *Cummerbunds*. Were midriff belts or sashes, of plain or pleated cloth, worn with dinner suits and bow ties, usually black, but also different colours to show colonial affiliations. From an Indian word.

9. *Aburi Botanic Gardens*. In the Akwapim hills built in 1875. Later used by Administrative Officers in need of recuperation. Opened as the Botanical Gardens in 1890, with links to Kew Gardens.

10. *Korle Bu*. Was an African Hospital of 221 beds, begun in 1920 as part of Sir Frederick Guggisberg's development plans, and completed in 1923. It had electric light, a sewage system, and fine grounds.

11. *For health and practical advice*. "Hints on outfits and the preservation of health, with practical advice for Colonial Administrators". Ref. Gold Coast Diaries pp.381–4.

12. *Asuansi*. In 1929–30 29,646 plants issued to the public of which 28,851 were economic and fruit trees. Ref. Department Of Agriculture Report 1929–30 p.21.99.

13. *Jacob*. When clearing the late Jemima Lynn's flat in 1976 a faded photo of Jacob was found on her bedside table, amongst family photographs.

14. *Hyrax-Dendrohyrax Dorsalis*. Is a rodent-like mammal of the tropical forest 17–24" long and 10–12" at the shoulder, 2–11 lbs in weight (6.6 lbs average), which forages at night on leaves, fruits, bark, twigs and grass. Well- known for loud piercing contact calls when foraging. Lives solitarily or in small groups, and is closely related to the elephant as it has similar feet and long tusk-like teeth.

15. *The railway*. Sekondi to Kumasi built 1898–1904 and from Accra to Kumasi was built from 1909–17. By 1920, there were 250 miles of track open in the Colony. For details of all travel etc. Ref. The Gold Coast 1931 (Census) Cardinall.

16. *Roads*. Lieutenant Colonel Watherston laid down a policy for road development in the Northern Territories (NTs), anticipating

the requirements of motor traffic in the Protectorate, and thereby setting into force a scheme which the first Chief Commissioner Northcott had suggested in 1899. Ref. Census p.36. The first roads were 14 ft wide and subsidiary roads 8 ft, with good ditching and raised roadways. Culverts were made of stone, where possible, or drifted. Ref. GCD P371. Bush paths were 18 inches wide. They were made and maintained by forced labour, with most work done after the rainy reason. The first mail sent by road was in November 1915 to Tamale, saving two days each way. 1914 saw the first motorcycle and sidecar in the North. In a letter to the Honorable Ormsby Gore, Under-Secretary of State for the Colonies, in October 1925, the DC notes there was a motorable road from Tamale to Navaro (104 miles) dependent on the Pwalagu crossing, closed in the rains from end of May to end of November. It had branch roads to Bawku and Lawra. "This makes a total mileage of dry season motorable roads within the district of 78 miles." Ref. CUL Fortes Papers. Add. Ms. 8405.

Chapter 2 - 1st Tour 1930–31

17. *Volta River at Yeji.* Used as a port for Gambaga, and extended up the White Volta to Port Tamale. Ref. A Historical Geography of Ghana, K.B.Dickson p.238; Letters p.52.

18. *The Northern Territories.* "Are 30,600 sq miles and are run by a Chief Commissioner. There are 21 DCs – 22 white men running a country with a native population of 530,000. Half a company of West African Frontier Force and two British Officers, 2 NCOs, 450 Northern Territories Constabulary, with four British Officers = 30 white men to keep overall control." Written in 1925. Ref. Letters, p.142. Cardinall in his Census of 1931 notes numbers of non-African population in the NTs as 1911 (13); 1921(36); 1931 (107). Ref. Census, p. 254.

19. *Calling cards.* This practice was a colonial protocol. It was customary to introduce yourself when arriving in any new station. It was usually a formality to leave cards in the afternoons, particularly on Heads of Departments and their wives – one

for a bachelor and two for a couple. Some had books to be signed, especially the Governor or very senior Officials. This introduction often led to an invitation to drinks, lunch etc.

20. *Growth of Tamale*. It was the decision by the British in 1907 to make Tamale its military and administrative HQ, rather than Gambaga. Tamale then was a Dagomba shrine village, but its central position led to its development. In 1937, 60% of the population were Dagomba, mainly peasant farmers; 24% non-Dagomba natives of the Protectorate and neighbouring French Territory, working as Waged labourers; 8% Nigerian Yoruba mainly market traders; and 8% Southerners mainly Government workers. Ref. Tamale notes in Fortes papers CUL. Historical Geography p.264; NTs of the GC. pp.362 & 371 Watherston.

21. *The Lions of the North*. Were the group of Colonial Officials who shaped the North from the formation of the Protectorate, at the turn of the twentieth century. Lieutenant-Colonel Henry Northcott, A.E. Watherston, H.T.C. Wheeler, C.H.Armitage, S.D.Nash, A. L. Castellain and G.B. Freeman were still greatly remembered in 1930; A. H. C. Walker Leigh, A. W. Cardinall and R. S. Rattray were leaving as Charles arrived, and P.F. Whittall, Dr P. Oakley, A.C. Duncan-Johnstone and E.O. Rake were his senior colleagues. For more details on the above see Ref. GCD p.385.

22. *Education facilities*. In 1909, a large Government primary school was started at Tamale, with a Technical and Agricultural School in Accra, taking a few students from the North. Other schools were begun in Gambaga in 1912, Wa 1917, Lawra 1919, Yendi 1922 and Salaga 1923. Formal education was introduced as a new concept, and part of the reluctance was loss of child labour for herding of animals etc. In the Northern Territories the Missionaries played an important role in education, and payment of fees could be made in kind. Later in Tamale a large secondary school with boarding facilities took children from distant villages. Ref. Census p.190. Boys only were taught in the early years. Schools would have an agricultural component added as part of the curriculum, and they each had a vegetable garden.

23. *Dr Phillip Oakley* was Senior Medical Officer Tamale 1930, on

the point of leaving to become Director of Medical Services Sierra Leone. Ref. GCD P393. Medical Statistics for the NTs in 1931 showed that for every 100 babies born alive, 46.3 died before puberty. Ref. Census p.217.

24. *Al Hajji Almajiri Mahamadu.* The Hausa tribe lived in the stranger's quarter or zongo of the town, and were great traders, along with the Moshi. In early Colonial times Hausas were brought in from Nigeria, and were the mainstay of the armed forces and police. The Mallams taught Islam and the Koran in the Arabic script. Ref. The Hausa Factor in West African History p.59 M. Adamu. After going on the Hajj on foot, pilgrims often brought new varieties of seeds home with them. Charles, who chose Hausa as his local language, was fascinated by the trade routes, and had hoped to cross the Sahara himself. He was very interested in stories such as that of Captain L. Binger, a Frenchman who in Salaga in 1888 met three Alhaji and wrote, "….They have through their travels also acquired some knowledge of geography which other blacks do not have." He cites countries in Europe and the Middle East. "Every good Muslim keeps a route map in a leather pouch showing the route from his country to Mecca..." "At a minimum, the pilgrims take seven years to make the return journey. They do not travel fast and are often obliged to work en route in order to earn the means with which to continue their journey." Ref. Du Niger au Golfe de Guinee pp.86–7. One of the main routes from the Gold Coast went via Timbuktu, Ghadames and Kairouan.

25. *Locusta Migratoria.* First swarmed in Africa around 1890. In May 1934 Fortes noted that the Chief of Tongo had no doubts that the rains would arrive, but feared the return of the locusts very greatly. For most disasters such as famine and drought there was magical cures, but they had none for locusts. The DCs had asked them all to turn out to beat the hoppers back, and they had done so in their thousands. Despite this the locusts still returned in swarms, in later years, to eat their crops. "Another instance of a pressing need without a specific magical counter-weapon". Anxiety was revealed by the amount of talk about their return, asking again and again if the white man did not have the means to eradicate them permanently

William Cook, a keen entomologist, graduated from Cambridge with Honours, and had contributed two scientific papers on locusts to the 1929 Agricultural Year book. He died in 1930 of arsenic poisoning, on locust patrol. He and Lloyd Williams were sent north when needed, but were normally stationed on the Coast. Ref. Department of Agriculture Report 1929–30 p.13, 19. GCD p.295.

26. *Chief of Walwale.* This demotion and "destooling" of the Chief was a perfect example of direct rule; carriers were essential for Government Officials needing to get to the smaller settlements of the area. The individual power of the DC to run his own district was important, as schedules of trekking were necessarily drawn up in advance, and Chiefs alerted to get ready rest houses with water and firewood.

27. *Nayiri, Paramount Chief of Mamprusi.* Captain Cecil Armitage, Chief Commissioner NTs, had the Na agree a document on his election on 14th April 1912, securing his recognition of British rule. This was the start of direct rule, with the aim of building up a Native Administration in the district. The natives of the Zuarungu district had been passive resisters, as far as obeying the chiefs was concerned. Heads of families, if sufficiently powerful, were a law unto themselves. The Nayiri was the head of the Mamprusi incomers, who arrived some 200 years before colonial rule, and now played an important functional part of the social system. They were distinct from the original indigenous culture, but married and integrated with them. It was the subtle, but very important differences from settlement to settlement, that so confounded the Colonial Officials as they tried to fulfil the remit of a unified integrated administrative entity. Ref.The Establishment of British.. Administration In Mamprugu 1898-1937. A.A. Iliasu; For History of the rise of Dagomba, Mamprusi and Moshi Kingdoms. Ref. Natives pp.5–11.

28. *Hammocks.* Were usually used on trek by DCs in the early years, when roads and paths were poor or climate difficult; sometimes also used across shallow rivers, so the officer did not get his trousers wet! Seldom used by 1930s. Fortes and Syme were carried out in a hammock, when too ill to ride or walk, and roads

were inaccessible. Ref.GCD p.246. Exceptionally a sick African woman was taken by hammock to hospital. GCD p.351. Also Ref. GCD pp.366, 367.

29. *Savelugu* was the seat of an important Dagomba Chief, reunited with the Yendi branch in 1918, when German Togoland was mandated to Britain. The Dagomba had originally settled in Yendi, but spread out to include the Tamale area, suffering enforced separation from their kin by the creation of German Togoland. It was noted that the links between the Nas of Dagomba, Mamprusi and Mossi were family ties, "and to this day observe the relationship by an interchange of messengers on all occasions of great national importance." Ref. Natives p.9.

30. *The Northern Boundary.* Was the Anglo-French Border mainly following the 11th parallel. The Intelligence Department made a very detailed map of the north. They were made up from "compass sketches made by officers and astronomical observations by the late Mr Ferguson, the whole fitted onto the rigid transverse work connected with the three Boundary Commissions; and on the whole a very accurate map has been produced." Ref. The NTs of the GC. Watherston; Census p.34. It was a great help to Charles doing his survey.

31. *Rest houses.* In the early days before rest houses were built, DC sometimes had to use tents. Ref. GCD p.246. In the 1930s some were still only simple grass houses, and their quality differed considerably from place to place. They were maintained by, and the responsibility of, each Chief. DC's demands on arrival were usually quickly met.

32. *Communal fishing.* Controlled by the Chief, who called the people to such festivals. They were important ceremonial functions.

33. *Bush paths.* Were about 18 inches wide, just tracks where carriers walked in single file, and in the rainy season tall grass often blocked the way. They were common property and everyone had a right of way through cultivated fields, and could drive their cattle through to water or to the grazing beyond. However, it was customary to make sure that there was no damage to crops; violation of this customary law was a source

of great local trouble. The country was thickly inhabited, but every man knew his boundaries and could point them out.

34. *Harmattan.* Wind from the Sahara, low in humidity bringing dust storms; caused by changes in weather from very hot to cooler temperature, especially at night.

35. *Funeral customs.* Fortes had a special interest in funerals, and describes these customs well, with the sacrificial ceremonies around a "swish" mound or Bagre. Rake in 1919, while DC in Tumu, recalls how the natives react to celebrating a funeral. "Yesterday everybody was very busy on their farms, today they have downed tools without a thought of their farming prospects and they will not take them up again for 3 days when the funeral rites are finished." He noted that the locals made the occasion a "funeral holiday", celebrating with feasting, drumming, music and a type of 'howling'. This was not only for a person of importance, but generally observed. "This sort of thing happens daily all over the district and might possibly be one of the causes of the annual hunger, which is always complained of at this time of year." This very situation was discussed between Charles and Fortes, and caused Sonia Fortes to remark sarcastically that Agricultural Officers would like to ban celebrations. Burial customs varied from settlement to settlement and were a very important ceremony to ancestor worshiping tribes. Ref. GCD P300; Tribes pp.184–214; Clanship p.121.

36. *Crocodile taboos.* Come under "taboos of the earth" and relate to animal totems. Some of these taboos are common throughout the area. Some are specific to certain localities and fall under the ritual jurisdiction of particular clans or lineages. A crocodile is sacred in specific places, often tied to important earth shrines, but in the bush they are not sacred. Fortes notes that amongst the Tallensi, "the commonest and most widely respected totem or quasi-totem animals are what the Tallensi call 'teeth-bearers' – reptiles and carnivores, whose weapons are their teeth who live and defend themselves by attacking other animals or even men." He noted in this a symbolic link to the aggression that an ancestor could cause an offending descendent. The animals are seen like 'people of the earth', and a symbol therefore of its mystical power, and the way it acts on people.Ref. Clanship.

pp.142–5. The Chief of Navarro's totem was a crocodile. Ref. Natives pp.36, 39, 41.

37. *Local trade resumed* in many parts of the North with the coming of British rule, mainly transformed by peace and improved road transport. In the days before British rule none of the original tribes would consider venturing to Ashanti, for fear of their lives or enslavement. Much of the local trade was agricultural in nature, and Fortes remarked in his diaries that its encouragement would enhance agricultural development. But it was the importance of the prevention of famine and soil fertility that really concerned Charles in early 1934, as there was usually little surplus grain. With the situations prevailing at the time, it would have needed a much greater amount of work for relatively little increase in profit.

   The worldwide influenza in 1918 was spreading along the old trade routes, and further hastened their decline. One in 20 in the Gold Coast was to die of this flu. Tumu and Gambaga lost their importance as trading posts because the new roads and lorry system bypassed them, as the administration moved to Wa and Tamale respectively.

38. *Shea butter tree.* The tree is named Bulyrospermum Parkii after Mungo Park the explorer and Scottish surgeon, who came across the tree in the 1790s and named it after himself. He was travelling for the African Society of London. It is the commonest of trees and could be home to spirits according to local beliefs. Ref. Natives pp.32–3; Department of Agricultural Report 1932–3 p.16. 69; The Political Economy of Colonialism in Ghana. G.B. Kay pp.217–227.

39. *Agriculture.* In 1920, A.W. Cardinall wrote about ceremonies held in North Mamprusi for land and crop successes. Ref. Natives pp.82–94.In 1929, he noted, ".... the religious beliefs of the people play probably the most important part in agriculture... For five to six months in the dry season, no cultivation takes place, and the people live on foodstuffs stored after harvesting." Conical storage huts allowed provision for about 250 lbs of dry grain per head, to take them through this time.

   In the sandstone area of South Mamprusi, around Tamale, the system of cropping was based on bush fallowing; crops were grown for periods of from three to five years. The area was progressively abandoned to recuperate, for possibly 15 to 20 years. This depended on the density of population, and the inherent quality of the land. The main crop of the area was the yam (Dioscorea) grown from sets placed in mounds roughly 4 ft x 5 ft. These were well made with a big- bladed hoe, like a hand plough, which represented quite a considerable cultivation. There were many named varieties of yams, with varying times of maturity, keeping and cooking qualities.

The basic problem in the northern region was the maintenance of soil fertility, in circumstances of a reduced fallow period, and a growing population. Fertilisers could not provide an economic answer because of transport costs and the low value of the crops grown. Much of the work at the Tamale agricultural station was necessarily devoted to fertility trials, with green manure crops such as mucuna, crotalaria and pigeon peas to find an alternative to bush fallowing. The station was originally established in the early 1920s as the cotton investigation station, to replace an agricultural garden located on the European Residential Ridge.

Zuarungu Research Station was developed in 1932 for the different soil conditions – granite and greenstone – in North Mamprusi. A new monthly magazine was started in May 1931 called "The Gold Coast Farmer" and had immediately over 400 annual subscriptions.

40. *Fred Mackay* travelled out on the Aba to the Gold Coast in December 1933 with Meyer and Sonia Fortes. He had been DC Gambaga for three tours, and said that he knew Rattray well. He had joined the political service from the RAF, and assured Meyer Fortes that the officers in the north would be interested, and value his work. He even assured him "several times" that they would be grateful for his advice. "Referred to G.C. Government policy of re- crystallising 'native state' for indirect rule. This left to the imagination of the people with suggestions from DC all now keen on getting hang of Mamprusi." Ref. Fortes Diary 3rd January 1934.

41. *A Tendaana* was seen as a high-priest and custodian of the land of his tribe and ancestral spirits, of the original inhabitants. In Dagomba the word Tinga was the word for "land", the Tindana (sic) is the man who looks after the Tinga, and the people

lived under his magical protection, and he ruled their 'souls' The Chief's power,however, was from the Namoos community, and was the focal tension in the society. Most of the farmers and their traditions were governed therefore by the Tendaanas, but Charles was required to work through the Chiefs by the Administration. Rattray thought the Tendaanas would have evolved in time into "Chiefs", but were interrupted by external influences, i.e. by Mamprusi overlords who created their Chiefs, who had "naam", and who therefore had secular and physical power. A dual system of leadership arose in this way, with the Tendaanas being seen to own the land and the Chiefs owning the people. For the story of how Chiefs arose and were accepted by Tendaanas. Ref.Fortes in African Political Systems, p.245; Clanship p.181 and Religion P56. Amongst the Hill Tallensi Tendaanas, the Golibdaana's importance was for the Golib festival of Dances at sowing time, and the Bo'arana for the Bo'aram celebrated at the end of the harvest and onset of the dry season in November. This latter was the main collective sacrament to the ancestors, and was a time of ritual initiation into the boar (bo'ar) cult for young men.

Each particular area had a number of earth gods to guard it, and was under the control of the Tendaana. These gods abide in natural objects, such as clumps of trees, rocks of any large size or remarkable appearance, and ponds; but clumps of trees are their favourite home. The groups of trees called Tengani (Tongban) Groves, were of great religious significance and dotted the countryside. Appeasing the local earth god was of the utmost importance, for neglect would cause famine, and worse. Also, important shrines were found in market places, for protection. For any breaking of taboos atonement had to be made by the Tendaana, who was regarded as Owner of the Land, or Chief of the Earth. He also regulated the day when the new crops could be eaten by the community; in short, he regulated all matters that concerned the religion of the Earth-God. Soothsayers or diviners also played an important role. Ref. Census p.63; Clanship p.184. "Locusts are the palaver of Tendaanas. Locusts come from the earth – out of holes – for where else can they come from? All this, as well as the fact that they consume what grows on the earth, is why Tendaanas have to concern themselves with them." He noted that they prayed still to their own ancestral shrines for protection against them. Ref. Fortes Diary 8th May 1936.

When the new Court Tribunal structure was formed, Fortes' advice was taken, and as a result the Tendaanas were included as part of the dual power structure that was first emphasised by Rattray. In a Christmas letter in 1937 Kerr wrote to Fortes about changes in the district , especially in regards the Tongraana trying to make himself the "big boss of the Tallensi". In his reply in January, Fortes remarked he was not surprised, as all the main Fra-fra chiefs would have done the same. It was for this reason important that Tendaanas should have their correct place to keep the balance. As delegates to the court, they should be a representative of a smaller council, and so responsible to their own settlement and section. He noted that both the important chiefs, the Tongraana and Sieraana, could not have gained their positions unless they had the agreement of the tendana who owned the land on which they were "sitting". He alone could give them the proper religious rights and legitimacy. "If there is one thing that both of them fear, almost as much as they fear the white man, it is the curse of the Tendaanas; and I believe that these religious sanctions could be transformed into political sanctions by giving the Tendaanas, or their representatives, some say in the government of the people". Fortes then made a profound remark "The political education of a primitive tribe seems to me to be a bigger job than just administering them, and to demand a lot of knowledge of their character and other features of their way of living than can be seen on the surface or put down in official reports." Ref. Fortes letter to Kerr Add. 8405/2/1938/6 CUL. Tendaana power significantly changed, as Christianity and Islam also competed for adherents.

42. *Effect of entrenchment on the NTs.* In his 1931 Annual Report the Chief Commissioner, Major F.W.F. Jackson noted that due to retrenchment the post of Deputy Chief Commissioner and one Provincial Commissioner were abolished. The remaining Provincial Commissioner's Office was amalgamated with that of the Chief Commissioner. Indirect Rule was imminent and the country was to be administered directly by the Chief Commissioner. He had a staff consisting of a Provisional

Commissioner and fourteen District and Assistant District Commissioners, and eleven former districts became seven. The Ten Year Development Programme was thrown into question as the Colonial Government committed itself to a low level of taxation, and reduced the rate of export duty on cocoa after the slump. Ref. Political Economy P28.

43. *European Health.* Fevers, dysentery and boils were common. There was a definite change and improvement after the Second World War . For good statistics and information Ref. Census pp.262–5. Early deaths of important DCs in the North included Watherston in Tamale 1909, as he was about to go on leave. Captain Breckenridge, 1910 on the way home, Captain Swire, 1913 in Tumu, Wheeler, 1916 in Gambaga, A.J. Berney, 1917 in Gambaga, and Sheriff was invalided out in 1917. In 1898, for any length of stay over one year's service the Officer had to be examined by four doctors together, with a report about malarial microbes in the blood. Blackwater fever was a common and often fatal complication of malaria. For instructions to Medical Officers in 1899. Ref. GCD P371. Cardinall advocated segregation from the African population to avoid infection from malaria, yellow fever etc. Ref. Census p. 262.

Chapter 3 - 2nd Tour 1931–33

44. *Agricultural survey.* A 1925 letter from the DC Zuarungu to Honourable Ormsby Gore, Under -Secretary of State for the Colonies, records, "No scientific research work has been carried out in the District at present." By the mid- 1930s it was said, "It is one of the few areas in West Africa where a really sound agricultural survey has been carried out as a basis for agricultural improvement. The result has been that considerable success has been achieved within a comparatively short time, and the confidence of the population has been obtained." Mixed Farming was thought to be the way forward, and Nigeria was putting it into practise. However there they had not done a survey, and the conditions were somewhat different. It was in fact based on improving native practise. It was hoped that with

intense methods of farming a younger generation would stay home, rather than go south to sell their labour. Ref. The Political Economy. p.233 Kay.

45. *Fra-fra Country.* Annexed in 1898 Taleland in the 1930s was still a backwater of the Gold Coast . Bolgatanga-Zuarungu was becoming a new urban centre with the coming of all weather roads from the south. Tongo itself was a 2-3 hour walk, and most of the Tallensi area was still very underdeveloped. The DC's post in Zuarungu was sometimes abolished, due to retrenchment, and the area was deemed not to have received "the innovations associated with British Rule." Ref. The Economic Basis. p.196 Hart.

46. *Food shortage and diet.* Were age- old problems and a cause of concern in the early years of British Rule. For famine in 1890s, Ref. "Food supply in Fra-Fra," by H. Wheeler CNEP, Accra NAG ADM 68/5/1 31st May 1911. Cardinall in 1920 recalls the kind of diet prevalent at the time. Ref. Natives P82, 3; Clanship, p.173. On continuity from the pre-colonial era into the 1900s in the practice of pawning children for food in times of scarcity, Ref. Zuarungu District Informal Diary 15th December 1913, by S.D. Nash Accra NAG ADM 68/5/1. "There is a great shortage of food this year, so much so that money is almost useless. I heard of yams being refused at the unheard of price in the country of 6d. Corn is also very scarce. The country here is as dry as a stick." 10th May 1918, Yendi DC Poole. Ref. GCD p.352; Agriculture In North Mamprusi, p.10 Lynn. Rattray employed Victor Aboya, from Winkogo, who had been sold as a "domestic slave" by his uncle in a time of famine, for a few baskets of grain. His name, Aboya, translated means "lost".

Fortes was told by Teroog in June 1936 that "hunger" times could be very hard. If they had to go to the Dagombas to buy grain, they would take sheep and goats to barter. As with the long distance trade caravans, they went well armed for fear of being grabbed and sold into slavery. They were pleased this had stopped now. It was so much on the minds of the people that the Kpanaarana, only a few days later, remarked that it was only the older generation who could remember these times. He concluded that only the British had brought this about. Lorries could bring up grain from the south, and it could be local or imported, as for the Nangodi Mine. But starvation was also in the power of the chief, as in the time when Tongraan Piribazra was taken to Gambaga and died. Immediately people started dying of hunger in paths and roadways. "As you went from place to place, you pushed corpses out of the way; they had no time and no strength to bury people, when you paid 5000–6000 or more cowries or an animal for a small cap of grain." They did not cook their usual millet porridge, but chewed each grain one at a time. "These stories, told with children sitting about, and are what tribal history exists here – the history of what living people of today remember." Ref. Fortes Diary. Further Ref. Long Term Patterns of Seasonal Hunger and Nutrition and Economic Destitution in Northern Ghana. Destombes; GCD pp.331, 332, 336.

47. *Enforced resettlement* near Salaga. No tradition of willing southward migration , for any reason, for this overpopulated group in pre-British days. From the beginning the great importance of ancestor worship, with ancestral shrines sited over graves just outside compound walls, was not considered important enough. Also, the soils favoured yam growing, while the settlers were by tradition grain growers.

48. *Sir Ransford Slater* acted as Governor for a few months in 1919, and again from July 1927 to April 1932. Christiansborg Castle was built on a site originally occupied by Swedes, then the Dutch, and finally the Danes. It gradually developed into a major building, but was severely damaged in an earthquake in 1872, 12 years after the British bought it from the Danes. It eventually became Government House and was substantially rebuilt. Ref. Trade Castles and Forts Lawrence. p.200.

49. *Lieutenant Colonel P.F. (Percy) Whittall DSO.* Was a military man who, like many others from the Gold Coast, served in the Boer War. He served first in Ashanti and later Tumu in a military role. He fought in the First World War and then returned in 1919 to be DC Tumu, and eventually became a Provincial Commissioner in 1928. Ref. GCD P300. Stewart Simpson the Vet told the story about a ghost called Ferguson (also said to be the ghost of Swire, who died and was buried there), a notable lion hunter,

who walked on moonlit nights through Tumu station with one or more lions at heel. DC Whittall had been aroused from bed one night by shouts of the Corporal of the Guard, ordering the night detachment to present arms. "Colonel Percy Whittall rose out of bed in some haste, pulled on his mosquito boots and with a hurricane lamp in hand strode across the compound demanding to know what in hell all the row was about. The corporal of the guard was quick to explain that Captain Ferguson and his lions had just passed by and, as always, the guard was turned out to salute him. This was too much for Percy Whittall. Next day he packed up his personal belongings, his office and his district treasury, and with carriers provided by the chief of Tumu he set off for the nearest provincial headquarters at Wa." That was the end of Tumu as a separate political station. Rattray never saw the ghost, but he told Simpson that the local inhabitants were convinced of its existence. Ref. Machin P168. GCD pp.330, 397. Allan Kerr said that Captain Swire had lived a solitary isolated life with his pet lion, and when he died he "walked" between the bungalow and his Office with his pet lion at his heels. The armed guard was always maintained at the district headquarters, but was seldom needed. It was such an eerie station that the story seemed quite credible and was widely believed. Percy Whittall wrote in great detail on Land tenure of the Fra-fra tribes assisted by Monsignor Morin and Father Barsalou of the White Fathers Mission. This had previously been attempted by Captain S.D. Nash in 1911, when he was Provincial Commissioner Northern Province.

50. *White Fathers Mission, Navrongo.* 1906 Bishop of Navro, Monseigneur Morin of Canada, established the Mission, extended to Bolga in 1925. The Governor offered Navro because a Government Officer was stationed there, so the Mission was under the government observation. All missionaries had to be British subjects and teach in English. Ref. Colonial control and provision of Education 1908–1951, Bening. They started with four priests, but no sisters. They also had Missions across the border in French Territory, Ouagadougou, Algiers, Tunis and elsewhere. Fortes states in his Diary on 27th August 1936, that Father Robert came to lunch with them, and they thought him "an eminently sensible man". "Taking for granted

his Christian Catholicism, as the firm and fixed groundwork of his life and attitude, one can appreciate his point of view and sympathise with his work here. To begin with, his open scorn of the administrative officer's ignorance of the people and country. They know nothing, he says. They believe in the chiefs who are one and all rapacious and extortionate." He was upset and "bitter" about the chiefs and elders, who all appeared to be anti the Mission. The former were hostile, and refused to convert to Christianity, and resisted efforts to convert their people. It was noted that from the Chief's and elders section of town there were fewer converts, for instance in the Bongo area. They even "oppress and persecute their sons who become Christians". Fortes noted that the Fathers were against the chiefs and elders, and this was one of the causes for friction between them and the administration.

51. *The Fra-fra* see themselves as consisting of three peoples, Grunsi, Nabdema and Tallensi, but they merge into each other without any definite boundaries. No apparent physical difference but dialect and culture (especially burial and totem customs) subtly different. The Tallensi, of all the tribes, celebrate the most ritual festivals.

52. *Role of farming in the NTs* was practically the sole occupation of the people, and even this work, especially after the introduction of the plough, required only a small number of the population at certain times of the year. In the 1930s, not less than 95% of the adult males were still totally dependent on agriculture for a living. After the Agricultural Survey the Chiefs and farmers in the areas selected got the most attention from the Agricultural Department, with DCs coming round about once a quarter on political duties. Ref. Farming according to custom; Food in the Domestic Economy.; pp.237, 242 M & S Fortes Africa 1936. GCD pp.330, 336.

53. *Indirect rule.* Sir Frederick Lugard's book on the "Dual Mandate" set out the aims particularly for the northern Islamic states in Nigeria. Guggisberg set the scene on the Coast, in the early 1920s, by the development of Korle Bu African Hospital and Achimota College. In 1932, a "Native Authority Ordinance" was introduced for the purpose of setting up local government

in the Protectorate, prior to indirect rule. The NTs were slow developing, so the changes in 1933 were more dramatic. The reorganisation of the system of government brought in taxation, the formation of the Native Authorities and local courts (Tribunals). Forced labour mainly became a thing of the past, exposing the personal demands for tribute by the Chiefs, and the role of the Kambonabas, their boss boys. Above all it sought to reinstate the authentic traditional leaders, the Tendaanas, who had been brushed aside in the early days of colonial rule. New positions of authority were created, providing wages for individuals who got the jobs through literacy or personal qualities; the old egalitarian traditional society and its customs of tribute etc. began to change, and was undermined by indirect rule. The Chief of Tongo and his elders complained to Fortes, all in agreement, that this arrangement was eroding the Chief's power. Nobody would now do the work they did before voluntarily. The ordinary people would simply refuse, and it was not possible to force them. In the early days a person could be hauled before the DC, which suited the Chiefs better. Whatever was needed would be done immediately, under forced labour, but this was now gone, and they preferred the past. "There is something ironical in this – a tradition-steeped chief like the Tongraana wishing for the sort of Direct Rule which ardent crusaders like Gibbs and reformers like Jones think so deplorable. Another example of the divergence between native wishes and interests, and what we think good for them." Ref. 11th May 1936 Fortes Diary.

54. *Pwalagu Drift.* Usually remade yearly when the water in the river receded in early November, and was a stone and stick causeway. Half the adult men in the area worked for 11 days when it was last repaired in 1936. (1, 250 men with 30 Kombenabas) £25 was distributed amongst the people in halfpennies or a tenth of a penny "anini"! Most of the workers at the time felt they got no benefit from remaking the road. Fortes, requested by Gibbs, also records what he found out subsequently about the Tongraana tribute. Every compound had to produce a working man or pay a fine to the chief's brother. A dozen of the strongest selected men were sent to work on 'home duty', were they would work assisting the chief and his close family -

mainly doing agricultural jobs. In 1934 they plaited zana mats for the chief, and were dashed a 3d each for 5–6 days work, and provided their own food. Ref. Fortes, Taxes and Tributes given to Tongraana in 1934 CUL.

55. *Zuarungu* formed originally as the garrison headquarters of the district, being close to the Tong Hills then very lawless. The market had a distinctive number of Mossi traders selling food, kola nuts and cloth etc. and a zongo. Zuarungu lost its administrative importance to Navro in 1930, then becoming primarily an Agricultural Station. For the colonial officers it had swish- mud houses with thatched roofs, and a small fort cum prison built in 1910 – with arrow heads embedded in the wall tops to stop intruders.

    Bolgatanga grew in importance due to new trade and its position on the main road, by-passing Zuarungu; as a consequence it was ravaged in 1918 by the flu pandemic. This preference made the Chief of Bolga very co-operative and supportive to the British. Bolga was to become the political headquarters in 1938.

56. *Thatching of rest houses* In December 1913, when building the new station, 2400 loads of grass and 110 sticks were brought in by the various chiefs for the roofing of the houses in Zuarungu; that was only half of the requirements. It was felt that this burden fell on the co operative chiefs more heavily, and those further away did very much less. Nash remarked that because very little money was paid for all necessary work, like road making and house building, it could "justly" be said to be equivalent to a tax. However in neighbouring French Territory, they did the same work and paid "a very substantial tax." Ref. GCD P331

    Fortes, however, found there was a hidden cost to the people and says that all the Chiefs do it to some degree. Tributes, of roofing grass, timber and mats, had to be made to the chief and family as a percentage of materials when repairing the rest houses. Likewise when labour is needed for rest houses or roads etc a certain number of workers are drafted to do some jobs for the chief's brothers or the chief himself. All of this is unaccountable. Fortes, Taxes and Tributes given to Tongrana in 1934. CUL.

57. *Zuarungu Doctor.* Zuarungu was created one of the medical research stations of the Colony under the Medical Research Scheme of 1923. Nothing, however, in the way of research had been done. In the report to Ormsby Gore in 1925 it stated that, while a medical officer was stationed permanently in Zuarungu, the area covered was so vast that it was impossible to make progress anywhere. "The result is that very little primitive sanitation has been done around headquarter stations, for the rest the natives themselves live in precisely the same foul, filthy and unsanitary hovels in which we found them" Ref. Report on North Mamprusi to Ormsby Gore 1925 by DC, Zuarungu. Fortes Papers CUL Add.8405. Box28.

After the doctor left for Navrongo in 1930 a dispenser, Mensah, was left in charge of Zuarungu. Fortes observed in 1934 that still no biological or medical research had been carried out amongst the Tallensi, or their nearest neighbours. The dispensary at Zuarungu was very limited. There were many common endemic diseases in the region, for which the natives had no reliable medicines themselves; they did, however, use an extensive range of plant remedies, and were very knowledgeable about them. They pointed them out to Charles when he was making his survey, noting special trees.

58. *Fra-fra houses* were one or more sets of round rooms made of swish, 11ft in diameter, joined together by outer swish walls about 7ft high. Interior party dividing walls are often low enough to climb over, or need a small step or tree ladder. One or more conical grain stores, with a removable cap of thatch, are found along with a courtyard for cattle and perhaps a stable room for any livestock. Ancestral shrines were also a prominent feature. Ref. Rattray, Tribes pp.247–254 for shrines pp.215–221, 328, 353, 357.

59. *Clothing.* The indigenous people, between the wars, still usually wore traditional dress with men in skins or a loincloth, and the women fresh bunches of leaves, but this tradition was slowly dying out. The Mamprusi only wore cloth, and thought nakedness beneath them. An early DC's diary refer to the Tallensi as "leaf- clad people"; he was concerned by the effect of fevers and pneumonia, especially in the 1918 flu epidemic,

in the "unclothed race". Another problem was complaints, in the courts, of theft of money from naked men in the market. Bows and arrows were still being carried for protection against wild animals, but were forbidden in the market place where they could be used against each other.

60. *The Grunsi* are often called "Nankanni" by the Europeans. They do not use or even recognise this name themselves. It is actually the name given to them by the Yulise, the tribe living to the west of them in Navrongo district.

61. *Making and use of string from fibre.* Cardinall tried in 1918 to encourage the manufacture, trade and sale of string in the market, a commodity necessary in every day life for the local people. He found young boys making string for bags, to be sold in the market; he tried to buy some of the string but was refused, then was offered it as a gift. Cardinall records: "He demanded a shilling, and the father explained – leastways, he called it an explanation – that they had never sold string before – the bags, yes, but what could the white man want string for? This was not an isolated case." Ref. Natives p.96.

62. *Dogs as food and sacrifice.* In Tale folk tales, dogs and cats have quasi- human status. Ref. Religion, Morality, and the Person pp.255–6 Fortes; Food in the Domestic economy p.249 M & S Fortes; Tongnaab p.186; Tribes Vol. 2.

63. *FAO* This refers to a brief return by Charles in 1962, for an assignment of six months. He visited the North, and met many old friends.

64. *Monotony.* As early as 1913 Nash pointed out, "monotony is the greatest drawback to an outstation in the NTs." Ref. GCD p.332. The rainy season must have been the most difficult time as no visitors passed through, due to poor roads and washed-away drifts.

65. *"The Rock did not speak."* One of the first Europeans to visit the shrine was Nurse D. Nanor (Zuarungu), his wife and a guide, in February 1915. Later in a sworn statement he wrote of his experience. The guide told him that while there was a "voice" it was not the voice of a man, "and that no one could understand what was being said except the Fetish Priest." He

was told that if someone wanted to ask something of the shrine they had to approach it naked. As they approached the grove they were asked to remove their clothing, which they declined. They eventually reached the shrine, where they were met by the Priest. In a letter from Dr Fortes to Acting DC Kerr on the matter, "But careful watching leads me to believe that the 'Voice' is provided by the Yiraana(the caretaker of the Yanii Tongnaab shrine), who is the ritual sacrificer, and that he, or more likely the man he sends to act as the 'voice', collects the spoil. BUT he or whoever brings the pilgrim, the introducer, shares the spoil." Ref. Religious Racket,Fortes; Tongnaab p.203. When Rattray asked through Aboya for a blessing on his work, offering a cow in return, "a long wail came from somewhere in the back of the cave. Everyone then clapped their hands with a kind of Masonic rhythm for the space of at least one minute." Ref. Tribes p.363; also Machin p.157.

66. *Land settlement scheme.* The Governor alone could give valid title to any land, under the "Land and Native Rights Ordinance 1931", for any big changes. Ref. Fra-fra Constitutional Development, Kerr in Fortes Papers CUL Add. MS8405. Native people moved locally and were granted land from Tendaanas and Chiefs, and this was not effected. In June 1936, the Chief Commissioner Jones wrote to tell Fortes that he proposes to make an up- to- date record of native law and custom, dealing also with Land Tenure. He asks for information on four points and ends by saying, "I will have your version of it interpreted to the chiefs, who if they agree as to its accuracy, will be asked to put their marks to it." noting that it would be of real help to the administrative officers and "greatly appreciated." Ref. Add MSS 8405/2/1936/71CUL; GCD P327; Culture Contact as a Dynamic Process pp. 41, 42Fortes.

67. *Tongo Rest House* was sited at the cross roads of the Tong Hills to Ba'ari (Bari) and Gorogo to Yamlag, and right amongst the native compounds. It was known locally as "the barracks", as it formed part of the site where the army was billeted in the pacification of the Tong Hills in 1911. One of the most influential Chiefs of the region, the Tongraana, had his home compound here. Fortes found on his arrival in 1934 that, while only six

miles from Zuarungu, it was not on a motorable road. The large main market of Bari was close, but the nearest European- style store or banking facilities was in Navrongo.

68. *DC's role* was crucial to the district as he held great power, and his personality was very important. They were often transferred for short times to fill vacancies for leave or illness, or ran two districts. DCs were judged by results of economic success (trade), peace and quiet in the district, advancement from direct rule to indirect rule, Legal and Court and Financial matters etc. The very implementation of indirect rule would mean the loss of power for the DC, in favour of the Native Authority. Under direct rule they worked through the Chiefs, who would often do things in the DC's name for personal advancement. Fortes reported to Gibbs and Kerr some of these acts. Their interpreters likewise were essential, but they wielded much power and influence, as in the case of Bassana Moshi/Grunshi. DCs could be quite competitive amongst themselves, and keen to improve their district, with a wide range of specialist skills and interests. With so few of them, each one had a high profile amongst the locals, and was individually judged by peers and locals alike, with nicknames to match. Whatever situation arose they had to take control, and find the solution no matter at what time of day or night. They travelled the district constantly, meeting the Chiefs and making numerous decisions They were staunchly loyal yet critical of each other, and aware of the perils each one faced, especially through sudden illness. Distances were great, and some places very isolated. There was much discussion from 1930 onwards, by other groups of officers, about what was seen as "the divine right of the DC". The creation of new departments challenged the belief that the political officers felt that they alone, of all the Departments, had to be in control. With officers covering a large area they, however, were on the spot to check work for all departments.

69. *Sanitary arrangements.* In the Zuarungu District sanitary habits were poor. The hut interiors were usually kept fairly clean, but the livestock yard inside the homestead had a build-up of manure. A refuse midden just outside the entrance added to breeding places for swarms of flies. Fortes remarks,

"The stench around a Tale homestead in the rainy season is enough to shake the hardiest anthropologist. These noisome accompaniments of home life do not trouble the natives, who look upon animal manure and human excrement as essential fertilisers for their farms." Ref. Clanship p.8. Fortes in his Diary recalls on 30th May 1934 that Vaughan (Sanitation Officer) and Mothersill (DC Navrongo) had a concerted effort to improve the sanitary conditions in the 'Navrongo Federation'. They recruited a selection of about a dozen boys, and gave them rudimentary sanitary training, in a hope of improvement. They were employed and paid by the Native Administration from its own revenues. Vaughn usually spent 4 days a fortnight on trek within a radius of 30–40 miles. Medically he usually treated people with yaws injections, worm medicine, coughs mixture etc. Yaws was very common and there were about 500 cases a month in this area. People came long distances to the hospital,from a radius of 30 miles of Navrongo. Ref. Fortes Diary, CUL. In the 1920s, a simple injection of Bismuth was found to cure yaws, a nasty illness of skin lesions that could lead to mental illness and death. Caught in childhood, it is linked to syphilis.

70. *Officials.* Officers who had been stationed on the Coast or Colony found life in the North very different. Policy was dictated from the South, but implemented in the very different circumstances of the North. In 1918, Cardinall remarked indignantly, "And it is a common saying, that NT Commissioners can find no work!" He refuted this, saying he found much more work in the North than in Tarkwa, his former posting. Ref. CGD 337. Good roads, the postal service, conveniences of every sort and proximity to recreational facilities diminished going further north.

71. *Black mistresses.* The Cape Colony had a long history of having a mixed race community, which produced some important individuals. It was common policy, in West Africa, that governed monthly maintenance payments for children of such relationships. DCs would be sent to ask the defaulting officer for back payments, which could be embarrassing if the man happened to be more senior. This is an area of colonial life that will perhaps never be truly written about accurately, as it is often covered up by lies and evasions. Clearly Charles

did know some people in this kind of relationship. There was a book written by Doreen Millar called "Black Mistresses." also Ref The Grand Slave Emporium, St Clair. for such relationships in the early trading post times.

72. *Lieutenant Colonel G.N. Gibbs MC* (1887–1969) Gumbo, as Gibbs was affectionately known, was DC Gambaga from 1933 for many years and became Acting Chief Commissioner, just before retiring in 1945. He retired to Penalonga, in the Highlands of Rhodesia, from which he would come on "trek" to see Charles, in Northern Rhodesia.

Duncan-Johnstone (b.1889) joined the service in 1913, and in 1936 was Acting Chief Commissioner of great repute in the NTs. He won renown for the handling of a smallpox outbreak in the North-Western Province in the 1920s, by making an isolation camp to stop the infection rate. Personal status was gained early on by being engaged to the niece of Lady Guggisberg. He also brought ex- King Prempeh back from the Seychelles after his exile. Harold Blair recalls him as being of a controlling nature, with set ideas of doing things.

73. *Attitudes to farming life.* Ref. Productive Efficiency and the Bounds set by Nature P173; Fortes; Religion, Morality and the Person, Fortes. Farms are protected from thieves by the use of magic medicine. Ref.Natives p.86. This point is well highlighted by Fortes. Ref. Clanship pp.8, 9.

74. *Reports and Diaries* kept by DCs and others. All officers had to keep detailed diaries of all work in the district. If a DC had to run two districts he would have to fill in two diaries. This was for continuity, necessary with constant change due to leave, illness or death. It was also a way of keeping in touch, within the Department, and would be read by the Governor. Handing over notes had to be detailed, and it was not uncommon for the Governors to write to the Chief Commissioner regarding an officers notes. Charles told the Fortes that every political officer also had to keep a confidential diary on their colleagues, in particular noting anything scandalous- real or imaginary. These types of diaries were misused sometimes, and "grossly libellous comments" passed up the chain of command to the Colonial Office itself. It was seen as an underhand way of

behaving, as people were condemned with no redress. Ref. Sonia Fortes Diary March 1934. In an article in the West African Review of June 1936 it was called "...the moist vicious single factor behind the demoralising of the Service." Its anonymity enabled some petty remarks to blight a career or even have an Officer dismissed. It broke the trust between Officers. The writer remarks that a service that allows an Officer to report secretly and adversely is "rank and rotten", and the reporting officer cowardly.

75. *Deadlock with "rain makers".* When drought or epidemic threatens, Chiefs and Tendaanas meet together to decide on ritual measures to be undertaken to avert it. This includes gifts to Tendaanas. Ref. Religion, morality and the Person p.183. It symbolically highlighted the roles of Tendaana and Chief, i.e. Spiritual vs. Secular power; Indigenous vs incoming protective overlord.

76. *Chief of Bongo. Anane Fra-Fra* worked closely with Charles and was an important part of his survey area. He had always been quite co -operative with the British, and was Chief of Bongo during the Riots, some 14 years previously. Castellain, Armitage and Wheeler were in political charge at the time of the Bongo Riots. Governor Clifford blamed the officers for not realising the seriousness of the situation, and acting before the tragedy got out of control. Ref. 1916 Bongo Riots, R. Thomas. GCD P 362.

In April 1916 there occurred a serious dispute over ownership of farmland between two families living at Bongo and Lungu. It was two and a half hours march for the Constabulary from Zuarungu. It started with names calling and stone throwing, and deteriorated into a fight with bows and arrows. Soon more people joined in till things got serious. The Chief of Bongo rode out to stop the fight, but when unable to do so rode to Zuarungu for help. All the young men of the area, who had been prevented from traditional fighting, now joined in the fray. The Corporal and his men from Zuarungu arrived just as the combined forces were about to attack Bongo. While trying to restore order the Corporal was shot by a poisoned arrow, and died a few hours later. Another Constable was wounded. Bongo was attacked and burnt and its people were driven into the neighbouring hills. This must have been a familiar scene reminiscent of the resolving of disputes etc. before British Rule.

The evidence given by the Chief of Zuarungu during the Enquiry is so illuminating and expressive of local native thought and feeling. While there was a Commissioner at Zouaragu (the earlier spelling) there was peace. The rumour went round "when the white man goes away, we will teach the Chiefs some sense." Without this surveillance shouting and then fighting broke out, causing the Bongo Chief trouble. "The people would not look to the white man who only travelled from another District. They said it was like a trader (white Hausa man) coming and going and no one would take notice of travelling strangers. I am sure that the people were not against the white man." They thought they could disobey the chief and not be punished, because the DC was no longer there to give direct visible support. The Chief Commissioner remarked "The old men have misled the young men recently. The old men don't care." Ref. Annual Report on the NTs of the GC for 1916 No 956 Accounts and Papers Vol XVII (1918) pp. 129–153Report No 970. Accounts and Papers Vol XVII (1918) pp.155–167 C.H. Armitage.CCNT

Acting Chief Commissioner Guthrie Hall in 1942 suggested the Chief was like an Albanian Brigand, for his extortion rackets! Ref. Bongo Riots; Roger Thomas. He was eventually destooled for murder.

77. *Communal work.* Forced labour was abolished in 1933, with the creation of the role of the Native Authorities to collect tax, and thereby pay cash for necessary roads and other facilities etc. However, locals did come together to do some projects for their own communities, with minimal external financial help, and great communal benefit.

78. *Attitude of Director.* From the formation of the Gold Coast, the Agriculture Department had lower status than any other department. Annual expenditure on agriculture was small in relation to the revenue of the country. There was general apathy towards agricultural problems in the North, cocoa being the revenue earner. Ref. The PoliticalEconomy p.199 Kay. See also Department of Agriculture Yearly Reports. No commercial crop was found to thrive easily in the NTs, and transportation

was very difficult and costly. Early work of the Department, using itinerant instruction, became insufficient and ineffective —it was seen that the lack of continuity could not solve complex problems of the Agricultural Department.

79. *Movement of labour.* Gold Coast Slavery Ordinance No 1 was passed in 1874, effecting domestic slavery. After the Ashanti Wars and the arrival of Pax Britannica, all slaves were freed. Those from the North, who knew where they were from, went home. Previously slave raiding etc. had prevented waged labour movement, but, with the loss of the slaves, waged labour was needed, and was essential for Ashanti in the 1920s and 30s, especially on cocoa plantations and gold mines. The North with its high population density, and low soil fertility, could provide the labour needs of the South. Ref. Economic Basis p.193 Hart.

80. *Reflections on social change.* Watherston, too, reflected that the benefit of consumerism, on self sufficient societies, may not be "a happy time for them." Ref. NTs of the Gold Coast p.360 Watherston.

81. *Eclipse of the Moon.* Rattray notes the names of the month, and the heavenly occurrences. The year begins in October and has 12 months. Tendaanas measure the days of the months with a piece of string tied with 30 knots, untying one daily. Ref. Tribes pp. 334–7 Sonia Fortes, while visiting Yinduri in March 1937, noted how the Tendaanas had turned the night sky into a hunting story, especially referring to Orion's Belt (the Hare, dog, and owner of dog) and the Morning Star (the hunter collecting the guts). The constellations did not have separate names, except for the Pleiades - which had two names.

82. *Survey was discontinued* in February 1933 when Charles was sent to Ejura. It was revived in early 1934 and extended to Navrongo, based at Tono.

83. *Ejura.* This house had been lived in by Rattray, and used by Charles at the weekends. Princess Marie Louise recalls how the way to the little bungalow was up a winding road, two thousand feet up. "I was conscious only of the supreme loveliness of the scene. The bungalow stands in an open space on the very edge of the scarp, looking right away over a

dense mass of forest trees, with a very faint outline of distant hills on the far horizon." In her time Captain Puckridge, who travelled with the party, had laid out the garden terraces, with flagged paths and flowers. He had been very proud of it, but it now looked neglected.She felt that it was to remind him of home. "How an Englishman will try to reproduce England all that is English, in fact, home, wherever he is, even in the heart of an Ashanti Forest". Ref. Letters p.50.

84. *The Zongo*, or Strangers Quarters. Sonia Fortes in February 1937 describes a visit to Yagaba, and the Zongo where the Hausas and Zabermas etc lived. It was on the edge of the town, with a market in the middle. Every family had its separate compound, with a high wall of zana grass mats. The houses were similar to the rest of the town, but they had swish or mud outside walls. The Hausas grew trees, and cared for them;each compound had its own courtyard well, dug by the owners, so they had nice clean water. The women wore clothes, were all very clean looking with smooth skin and red stained nails. The Zabermas were elegant dignified people, tall, pale skinned and respectful. As a group they seemed to be more culturally advanced, and many had Islamic names. During Ramadan some of the local people tried to fast, but soon abandoned it. The village women still seemed to go down to collect water from the river, and even their compounds had no fruit trees, while the Zongo was full of them. Ref. Sonia Fortes Diary; Tribes Preface X Rattray.

Chapter 4 - 3rd Tour 1933–35

85. *Economic botanist,* Trevor Lloyd Williams. "The Beetle" with J.C. Muir wrote "Agriculture in the Navrongo and Zuarungu District of the N.T.s' in 1930." Ref. Year Book 1930 Paper 27. Neither was stationed permanently in the North. Lloyd Williams died in 1942, of blackwater fever.

86. *Ancestor worship* and the belief in Spirits. The most important religious beliefs are in the power of the spirits of the earth, and of their ancestors to influence their fortunes. An often heard

refrain "This is the way our ancestors did it and we have received it from them." Before any serious decision or action was taken, e.g. a journey, sowing, harvesting or a lawsuit, a spirit or ancestor would ask for help. Misfortune means a spirit is displeased, and a soothsayer or diviner is consulted to seek the problem and ask forgiveness.

87. *Carriers* were still needed even in the 1940s, which gave local employment. The difference being they were now paid individually, but still organised by the Chiefs. The Chiefs, too, were paid a salary. Carriers had received compensation for food in the past, but the money was given to the Chief for distribution. Since 1921, with the development of roads, long distance head porterage of most goods was becoming a thing of the past, but still used between isolated compounds for officers on trek. Ref. Census p.150.

88. *Arrival of Fortes.* From his first field work report dated July 1934 it appears Fortes had expected to start work immediately upon arrival at Zuarungu or other Nankani area. Zuarungu was not suitable as the Residential Area was too segregated from the native compounds to be of much use. The rest house was also needed for Government officials on tour.

89. *On arrival in Accra Fortes* went to see the Director of Agriculture. He reports in his diary that he was "surprised at their keenness on research. Asked about Lynn's work in Zuarungu and saw Lynn's report. This was a strong point deciding in favour of starting there, as Lynn's report seems excellent basis for work." It was, however, to be in the political field that Fortes made the greatest contribution. Ref. Fortes Diary.

90. *Gibbs and Fortes.* Gibbs was reluctant to have Fortes in his district at first; Rattray, the first anthropologist, had not been universally popular. In the early 1930s all DCs in the Mamprusi Area were asked to write monographs, and they felt they understood "their people". Fortes with two years of diligent fieldwork, and the framework of "Structural Functionalism", added greatly to their knowledge of shifting allegiances. With the reorganisation of local Government in 1933 the timing was fortuitous. The social and political aspect of Fortes' study was to play an influential part in guiding DC Kerr in the Tallensi

reorganisation in respect of the Courts, Treasuries, the Kunaba, Tongraana etc.

91. *M. Fortes, A.W. Cardinall and R.S. Rattray.* Before leaving England, Fortes wrote to Cardinall in November 1933 asking for general advice, introductions to people, and suggestions as to where to base his fieldwork "a part which is not a complete backwash, but is of real interest and coming importance to the Administration so that our work will not be left to languish in the British Museum, but will be of use to those concerned with the development of Africa." Ref. Add. 8405/2/1933/48 CUL. He even got a friendly letter from Gibbs, through Cardinall's mediation. On the question of entertaining of DCs Cardinall says, "No one expects much, all they like is that you are self contained so that they are not worried about you – not by you. Things do not matter so much. I should not take luxuries especially for hospitality purposes." He then details requirements adding that, "gin is against the law. Whiskey you must have for entertainment, casual callers and so forth." Cardinall's moniker was Bugim di Kugri. (Fire has eaten the stone) For Biographical details. Ref. GCD 386. Cardinall and Rattray were in some sense in competition, as they both published books and articles around the same time. "The Gold Coast Record", set up by the Secretary for Native Affairs, was a house journal to which everybody was encouraged to contribute. Rattray contributed a sole article, and that was a re- hash of an article in Blackwood's Magazine. Cardinall also recorded folk tales, and was a successful DC. He saw Rattray as an upstart who wanted to make a profession out of what he did as a hobby. However, there was always a tolerance between the two officers. On 12th January 1929, Rattray flew over Navrongo in a thick harmattan haze, on his epic plane flight. "He scribbled a note to Cardinall and put it in a metal tube. As he circled the bungalow, Cardinall was waving to him and he dropped the tube." Ref. Machin p.64. Fortes in a letter to Professor Westermann (ex- missionary from Togoland) 2nd July 1935 remarks, "One of my disappointments was to discover how superficial Rattray's work was. He has a chapter on Tongo where we lived; and even the names were wrong. I traced his informants and used them, and was told by them that Rattray was often misled by his interpreter, who

was a Nankani." Fortes felt that in other areas too, Rattray had seen too much Ashanti custom into the tribal life of the Northern Territories. Noting the northern tribes were classical examples of ancestor worshippers, he planned to make a detailed study of their religious life in the second tour. Ref. Add. Mss8405/2/1935/41(i) CUL. Fortes was told how Rattray did his field work, and how he got one of his monikers "...and was known as Boardaan because he used to walk about and ask people about their Boares (shrines)".

92. *The Kambonaba* (Captain of the Gunman in Ashanti) were a class of people who emerged on the arrival of the white man, to represent local Chiefs and supervise enforced labour. In some cases their real agenda was to gain power, prestige and wealth. Often acting as interpreters and intermediaries, they became indispensable to the DCs where Chiefs seemed weak. Self confident and powerful they were a strong force, with powers to influence situations. They should have left office when the Chief died or was destooled, but in Taleland they tried to maintain their position of power. This led to some final funeral rites being postponed for years, if necessary. The strength and purpose of their role differed among the tribes.

93. *Nangodi mine.* Opened in 1934, and closed 1942. Sonia Fortes recorded the opening in her Diary from Zuarungu 9th February 1934. "Heard from Mr Lynn this afternoon that gold has been found at Nangodi; a place 12 miles away from here. Several lorries with machinery have passed here today, he says. He thinks it will be a great thing for agricultural work here, if a mine is opened at Nangodi, as it will absorb the surplus labour. Wonder if he is right." She feared that people would be drawn to the mines for the "ligir" (money) and easy means of sustenance, and therefore would not bother to improve their farms. On a visit to Nangodi they found the place most attractive and met both Mc Guinness and Reid, an ex bank clerk. She thought them hard men, as tough as nails. They arrived in the afternoon and found them sinking a shaft looking "grubby, scruffy bearded and grimy." They started with 40 men working for them, and had built a simple galvanised sheet shelter over the shaft. In August 1934 the Fortes were invited to stay with the Gibbs in

Gambaga. On the way back they met Mc Guinness at Pwalagu. He had a £2,000 bar of gold with him, the first produced of the mine. He was taking it south. Fortes recalled "He looked a complete buccaneer with his huge automatic and all." Mc Guinness took the opportunity to complain to Gibbs about the Government prohibiting all prospecting for gold except that of Gold Coast Selection Trust, claiming it "retarded" the economic development of the country. After this lecture "Gibbs was very cautious in his replies telling only that he ought to lay the matter before the Chief Commissioner."

The Fortes had to rely on the mine to get many conveniences, and were often asked to dinner, as in 8th April 1935. There was quite and international colony of miners by this time. "Mc Guinness and Reid, a South American Welshman, a young Englishman, son of a clergyman, ruddy, red-haired, blue eyed with a prize fighter's body, absolutely built for a colonial life of hard outdoor work, two young Italians, who have practically run away from conscription in Italy, and the temperamental hunting addict, the German Manthey. They live in grand style with two refrigerators, a loud speaker, a luxurious Vauxhall car, etc." Mc Guinness and Reid were bragging about the great blessing they had brought to the local people, and were making all sorts of comments about the life of the local Frafras. Reid said they employed 360 men that month, with the figures varying according to the season. In the sowing season most of them leave to attend their farms, but in the dry season they have thousands of applications for work. The local people do most of the above ground work, repairing buildings cutting firewood, repairing roads and bridges etc. (paid by Government) The underground work is left for the Moshi, Zabermas, and Kusasi workers. Reid said that the labour was about 80% from the Northern Territories, and most from around Nangodi itself. Their monthly wage bill was around £380 per month, most spent locally. A meal was given, paid for by the mine, to each shift, half way through their time - 3d for 10 men! Mc Guinness thought the nutrition of the workers had improved greatly, and they worked better as a result. The Veterinary Department agreed that there was many more cattle in the Nangodi area than the year before, as some of the money was going into buying cattle. In order to feed their employees the mine had imported food supplies against the approaching hunger season. They had ordered 7 tons from Kusasi and French Territory. "Every year, they say the country starves. This year there will be no starvation in the lean months because of the mine money going into the country." Ref. Sonia Fortes Diary.8th April 1935

94. *Local scenes* - Zuarungu Area. Fortes Diary recalls these scenes in Zuarungu District March 1934. After first arriving in Zuarungu the Fortes met Awabi, an old musician who came round every few months and played for a living. He always came to visit Charles, who used to give him a small dash. Later that day a group of women came to put down a new veranda floor of dung and mud. They each used a kind of mallet, with which they beat the floor in rhythmic time as they chatted and pounded away. In July 1936 the Fortes went to stay at the Kpataar, a simple Rest House made for them on the top of the Tong Hills. It was a very hilly area with the rest house nestled among the large boulders just on the edge of the cliff: the whole place seemed quite overgrown with vegetation since the clearing of inhabitants from the Tong Hills in 1911. The people then had been required to move and live in the valley below. In the quiet of the early morning the Fortes viewed the compounds dotted around beneath them in the valley , with their flat roofs just protruding out of the growing early millet. From this vantage point they could see small boys driving their herds of cattle, their little voices carried on the wind high-pitched and cheerful, as they went to seek pasture. In the evening they saw other villagers returning from cleaning and hoeing their millet, carrying their hoes and baskets. One typical family group caught their attention. There were some small girls playing with serrated calabash disks on sticks, then a small boy piping on a 3 note whistle and a bigger girl carrying a basket of grain. This was followed by a woman with a basketful of Shea nuts and other things from the 'bush' farm, and a man with a calabash on his head and carrying his hoe in the typical way, over his shoulder. What interested the Fortes was that they were members of the original tribe of the area, rather than Namoos. They noted their good naturedness and happy demeanour. That evening under a full moon, in the quiet of the day, the air was cool clear and the outline of the trees and rocks stood out in silhouette. The Fortes

enjoyed the necessary change and rest, but it was just such a scene that Charles found so rewarding too.

95. *Fortes transport.* For the first year of fieldwork the Fortes had only a horse and a bicycle. For the second tour they had a car. Despite the importance of the Tongraana the road was not motorable in 1934. Unless the DC came to Tongo, the Tongraana would be summoned to Zuarungu, coming in on his horse in the traditional way with musicians, drums etc. Charles taught both the Fortes to ride, and in the early days before the road was completed would drive part way to collect them, or drop them off. They would walk the rest of the way.

96. *Tingoli, Chief of Tenzugu* also known as Tengoli or Tengol. He always had style, and his stately bearing showed his high opinion of himself due his status as Chief, and his role in the Golib festival and important shrine. He was a tall imposing good looking man, full of self assertion and used to entertaining important visiting dignitaries from the south. In his own compound he wore the traditional dress of a loin cloth and adorned himself with a metal and leather necklace. On 10th May 1934 the Fortes decided to pay a visit to his hill top compound, only a few miles away. Sonia recalled that "No one else I know here could look more like a chief than he does, and Tenzugu seems the most suitable setting for him." Indeed Tenzugu and Tengoli seemed to really belong together. Only 4 compounds had been allowed to return to live on the hill after 1911, and this gave him a superior position, and fuelled the political feud between the Kpataar Tendana and the Chief of Tongo. Tengoli's compound was well situated, with cultivated farmland around it and some good trees. The rest of the area was wild and vast. His compound was an large imposing structure, freshly plastered, with many round houses with flat roofs on the perimeter as in the usual manner. However his own quarter was a large rectangular room, about 15ft high and built of sun dried brick. To do this he had employed boys who had been trained by the White Fathers at Bolga, but its decoration was in the Tallensi fashion. The view was excellent, especially the one looking towards Yinduur in the distance. The Fortes went to the edge of a rock, a promontory witha sheer precipice below, to see the countryside. The valley stretched

far in front of them, and was quite brown, and in the distance a small winding river. The flora of the plain below was termed 'park savannah' and from a distance resembled an orchard. The Fortes talked to Tengoli mainly of agriculture, his ancestry and the Gologo Dances. Tengoli was very proud of his family's traditions, and 'his' people, who lived in the valley below. He was sending three of his sons to the Gambaga School. One had gone to the Bolga School, but he was dis satisfied with his education, and felt Gambaga was superior. He said that this son even wore a cross, and he "approved" of him being a Christian: later, wistfully, he remarked that in his heart he had his own doubts about this, adding "If the white man says a thing is good, then we must agree that it is so and accept it." Add. Mss 8405 CUL.

Some two weeks later, by contrast, Tengoli paid a return visit to Fortes who had been home writing letters all morning. It was afternoon when Tengoli appeared on a horse, well clothed and in all his finery. His drummers and retainers came with him, including a Mallam from Savelugu. He was asked into the house and offered a cigarette, which he smoked with 'relish'. He confessed that it was his first, and then tried to imitate Fortes style by blowing the smoke out through his nose. His entourage was very amused! A picture of Tengoli, taken by Rattray some five years earlier, was produced much to his amazement. As they had no mirrors then, he did not even recognise himself, but remembered meeting Rattray. The event seemed to have a powerful effect on the Fortes "I think now how extraordinarily common place it seems to be talking to a man whose picture I was familiar with 2 years ago, whom I hardly expected to meet, and who was a piece of literature. It is as if an ancient document was to arrive alive unexpectedly – so unexpectedly as to leave us no time for reflection about the miracle, or for working up the attitude of romantic awe." Ref. Fortes Diaries 22nd March 1934 Add. Mss 8405.CUP. Tengoli died in 1949, his rivalry with Tongraana ending with Na Bion's death in 1941; whereon Tengoli managed to get his preferred man made Chief of Tongo.

97. *Lynn and Fortes.* Charles admired Fortes for his understanding and intellectual analysis of Tale society. Both were young men

at the start of their careers, Charles was 25 and Meyer Fortes 27. They supported each other in many ways, as each was for the first time putting his skills and knowledge to the test. Charles referred to Fortes's work as 'writing social algebra'. This was understandable as one of his aims was to create a theory of Social Anthropology, as encouraged by Professor Malinowski. On 17th February 1934, Fortes notes in his Diaries, "Later I had a talk with Lynn about his troubles. He is having difficulty in pushing the people to adopt the sort of agriculture and methods of crops which he desires them to do. But he has worked through Chiefs always trying to persuade them to adopt better crops and methods, in the hope that they will pass their experience on. I suggested that this might be the wrong procedure, in spite of the fact that it is insisted on by the political branch with its new enthusiasm for native administration. I suggested a) That he would do better to work through heads of section. b) That the administrative institutions were not necessarily fit to transmit and put through a change in economic institutions. c) That the business of passing on agricultural information and spreading through the people should be in the hands of the political department, not his, as they ought to be familiar with the necessary (existing) techniques."

Fortes in his forward in "The Dynamics of Clanship" (Xiii) records, "Mr Lynn's friendship was a great stand-by to us at all times….Farming is a subject of supreme interest to the natives; and though it does not come into this book very much, I should have made little progress in the field without some knowledge of Tale agriculture. I received great help from Mr Lynn on this side of my field-work."

98. *Social obligations.* When people came on tour it became a chance to get together and talk about current topics, and what was happening in different parts of the Country. Hospitality given was also repaid at some stage later. To the newly arrived Fortes it seemed that Zuarungu, despite its seeming isolation, was like a "'Tourist headquarters". This did not happen at all in the rainy season. In a place with no other social events, entertaining could be an expensive obligation. At this time Fortes noted Charles had to send frequently to Tamale for supplies of beer, which came up on the Government lorry. With the Fortes occupying

the rest house, and Gibbs in Zuarungu as replacement DC, Charles went on trek and allowed Mr and Mrs Hugh Thomas to occupy his bungalow. They always tried to accommodate each other where possible. The Thomas' were interested to meet Fortes, and a dinner invitation was forthcoming, including beer and wine. As head of Native Affairs in Ashanti, Thomas came up to see the progress of reform to the Native Administration in the Protectorate. Fortes saw Thomas is a very sensible man "without a trace of 'side' "

99. *Aeroplanes.* This must have been one of the first. Rattray made the first solo flight to Kumasi via Tamale and Accra in January 1929. Nigeria had the first Airmail Postal Service.

100. *Captain L.J. Mothersill* was an ex Sandhurst officer, who's career had been in India, Aden and Egypt. Fortes recalls meeting him at Christmas 1934 in Gambaga, where he explained that a military career no longer attracted him, as he found it a dull and narrow life in comparison to being a district officer. He chose the job because it gave him more opportunities and a better life style. Fortes recalls he had a quick mind and a determined personality, and took his work and life very seriously: he put his whole effort into what he did.

101. *Locusts.* On the 15th June 1934 a huge swarm of locusts swept into the valley from the Wakyi direction, attacking it first. As they drifted along there were some dense patches within the swarm. They moved across the valley in a north east direction, their numbers growing all the time. They passed over the fields and over the Tongo Rest House like a thick black cloud, wave upon wave they came making a great noise. Sonia Fortes recalls her "pain", as the corn was growing very well and was soon due for reaping. These were the hazards for agriculture, and she questioned the lack of scientific advance to prevent them. She reflected that they had attacked Natal in South Africa, an area where Englishmen had farms. "I suppose the officials do get tired of telling the natives to dig pits and destroy; but it is really expected that the natives will see the logic of a scientific procedure?" The people all ran out into their own fields, beating them off with sticks, shouting making fires and beating tin cans - anything to make the most noise. Charles told the people that

the best way was to cut grass and make a fire. Locust Patrols had shown the people what to do on many occasions in the past, and it was felt that they should now be able do it on their own account in future. It is true that locusts only arrived in Africa south of the Sahara from1880 onwards, and travelled down to Natal. Ref. The Scramble for Africa. P489, P41 Thomas Pakenham.

102. *A.F. Kerr. (1910– 2001)* Allan Kerr had a great influence on the politics of the Zuarungu District, and wrote the "Frafra Constitutional Development", after the advice of Fortes. Bolgatanga became the Political Headquarters in 1938, with Kerr helping planning its layout, on a grid pattern.

103. *Jim Syme,* DC Bawku and later Salaga. Sonia Fortes records in late November 1934, the events surrounding an incident in Bawku. Situated close to the border, it was common for men to slip across the frontier and out of reach of the law. So Syme went to arrest a man in the early hours of the morning, while still dark. He had been staying at a rest house close by with a party of policemen. The man shot a poisoned arrow, which lodged in Syme's lung,and made his escape. A policeman it was said had used a pen knife to cut out the poisoned flesh. After spending the next night at the rest house he left for Bawku, where Dr. Purcell looked after him. Apparently they had not been on speaking terms during the wet season. In July 1928 they had been stationed together in Yendi, and by and large got on well together. Purcell was not an easy person to get on with by all accounts, and sent inconsistent bulletins on his progress. Gibbs and the CCNT, Kibi Jones, brought a medical consultant from Tamale and he was diagnosed with severe pleurisy and in great pain. They operated at once and drew off a gallon of fluid and put him on morphine. But for this swift action Syme could have died. The culprit escaped across the border, and Gibbs wanted redress. However extradition was difficult. Ref.For hunting with poisoned arrows. Ref. Tribes,pp.1717, 175; Natives, p.47; Bad Arrows, Fortes. An incident not too dissimilar to this happened to A.W. Cardinall when he was DC in Zuarungu in 1919. Akanyele of Bolgatanga fired poisoned arrows at him, when the police came to arrest

him. A Constable died from a poisoned arrow and the episode ended in a shoot out and Akanyele's death.

104. *Death of Nangodi engineer.* Six months after the mine opened Mr Quonce died in Tamale Hospital, ten days after a rock fall that fractured his pelvis. He was nearly 60, and many though that was too old to be working underground. DC Kerr and Charles suspected that the safety of the mine was not as good as it should have been, and was never inspected. Charles recently had taken the Fortes to visit the mine, and on the way home they wondered if Quonce would ever see his home in Cornwall again, or leave his bones in Nangodi.

105. *"Kibi" Sir William Jones.* (W.J.A.) Chief Commissioner and Mrs Jones came to Zuarungu to talk to the Chiefs about indirect rule, on the 24th November 1934 where he addressed the Conference. There was a large gathering and with great ceremony he used the faithful Bassana as interpreter. He addressed them as 'chiefs and people', and went on to explain the new policy of indirect rule. He talked of the value of education, and the importance of learning to read and write. He emphasised that it was necessary to have chiefs who were also literate. He went on to explain how schools, wells and better medical facilities would come from this new policy. Of special importance to the Tallensi, he announced they could return to their old ancestral homes on the Hill.Fortes was an unseen hand here demanding that they all go back together, rather than in separate phases as the administration contemplated.

106. *Nominal Poll Tax for Native Authorities.* Fortes recorded in his Diary 8th May 1936, "Meanwhile the fiscal side of the new role is making progress by leaps and bounds and is the one pride of the whole administration from the Chief Commissioner downwards." The first full tax collection in 1936 was 1/- a compound for messenger money, which when collected reached around £1,800 for Mamprusi Division alone. Since this had proved successful, it was decided to increase it to 1/- a head in Taleland for every able bodied male. The heads of the compound were now called upon to collect the money, and hand it over to the family heads, who handed it over to the new Native Authority. The expected revenues from this new tax would be in the region

of £2000 -£3000. Nominal rolls of included males now had to be made to prevent tax avoidance. To a people totally unused to such a concept it caused dismay and anxiety, for it was difficult for them to understand the reasoning behind it. To them its imposition was pure and simply for the white man's purposes. Where would they get the money? Every head of family viewed his own situation, counting the cost. When was somebody eligible to pay, and what would happen to those working away from home? To a people used to living on the edge of hunger it seemed a big burden. Increased agricultural production had to be a big part of the solution. The concept of the provision of social services, as viewed by the administration, was difficult for them to comprehend. It was difficult for Chief and commoner alike to see the benefits. In many parts of Africa this would have been a very small amount to pay in tax, but given the level of the agrarian economy of the area it represented many more hours of work and effort. The European officers and traders all paid heavy import tax etc . Since it was of such importance, and a big topic of conversation to his field work group, Fortes tried to work out what impact the tax would have on them given their present economic resources. "I conclude that it is not too heavy; but the people are utterly mystified by it, and believe that the white man is imposing it for his own ends." Ref.Fieldwork Report, Fortes Papers Add. Mss 8405; The Economic Basis p.193. K. Hart.

107. *Climate.* The climate in March was very hot, and the Fortes found it especially difficult, as they were trying to finish all necessary fieldwork and outstanding information for the Provincial Administration. The Fortes left Tongo on 7th April 1935 having arrived in January 1934.

108. *Gorogo* settlement was one of Charles's original survey sites, so in passing he often stopped to look in on the Fortes in Tongo. The Chief of Gorogo was an elderly man who had a real interest in agriculture, and worked well with Charles. He was a firm favourite because of his passion for farming! Soon after arrival in April 1934 Fortes rode out with Samane, the Chief of Tongo's brother, to call on the chiefs of Gorogo and Sii. Fortes wanted to see how 'the return' was going, and watched

the laying of the foundations of Dievol's (an elders) house in Gorogo. The return of the Hill Tallis from exile was an important event for them, but it took them away from their more fertile fields. As he arrived he noticed a stream of men coming up the hill to help in the rebuilding. Many important people came or sent their sons to help. The Chief of Gorogo brought out some pito, and soon a festive atmosphere occurred. People came to greet Dievol, and one man said that he would long be remembered by later generations as the first man to return to rebuild his father's daboog, or ancestral home. To this profound greeting the reply came "tare, tare". The area in which they were rebuilding was on the slope of the hill, with huge scattered boulders. The houses were built in the open spaces, which had the odd sparse tree. They were leaving their home half a mile away down on the flatlands below, to which they had gone some twenty years before. Fortes teased them saying the flatlands were better. They would not hear of this, saying the homesteads of their forefathers were superior. "They have always said we must come back, have killed our children because of our throwing them away." They gave all kinds of reasons for this, but none of them seemed really logical, but then many people make illogical rationalisations because of sentiment about their home place. The will to return seemed a strongly held belief, bound up with magical sanctions. Sites were being cleared and everybody knew their old home ruins, most had no upright lumps of swish showing through the weeds. But one, belonging to an elder, had part of the wall of his zong still in place. It had been somehow kept in repair, and sacrifices had still been done on that spot during the years of exile. By tradition a bagare cannot be moved. On attempting to go into the zong area, the son of the elder begged Fortes not to break the taboo and enter, as he was not initiated. It was very clear that building was going on speedily and people were rushing back as far as from Dagomba. It was realised by the administration that there was a yearning by the local people to go back to their ancestral homes in the Tong Hills. The reason for their removal had long since passed, and a new generation had taken over. Gibbs was a very cautious man and advocated a phased return, to which "Kibi" Jones the CCNT agreed. They came to discuss this problem with Fortes, the day before an important speech to announce their decision. Fortes was aghast, and explained to them that by doing this they would run counter to the egalitarian society of the people themselves. The Commissioners might think it was an orderly return, but to the people themselves it would simply be viewed as preferential treatment of one group over another. Needless to say the Gologodaana was not happy, as he had been allowed to live on the hill. The result would be to spread hostilities and jealousies, and increase further underlying personal intimidation.

109. *The Earl of Plymouth* was a Conservative politician who held many important posts. He became Under Secretary of State for the Colonies from 1932 – 6, and Parliamentary Under Secretary of State for Foreign Affairs from 1936–9. As a result of this visit Jones received a secret circular, which he showed to Fortes, who recorded this on 3rd November 1936, "Jones gave me a secret dossier to look at. It was from the Secretary of State for the Colonies to the Governor about the Colony, urging inquiry into the reasons for the failure of Indirect Rule there, but holding to the formula that Indirect Rule through the Chiefs – the customary 'native authorities' – must be developed."

110. *Tumu*. In 1925, Princess Marie Louise visited Tumu and, having heard of its reputation for lions, was disappointed not to see any during her three day stay. She was told that, in the early morning of the day of her arrival, a woman had been carried off by a lion. She was thought to have been eaten by one as nothing was heard about her again. Ref. Letters P.101, and 104 for description of greetings they received at Tumu. Edward Bowdich, explorer, described a very similar colourful reception in 1819. Rattray complained of the swallows living in the rafters in 1929.

111. *Captain R.S. Rattray, CBE* was the first trained Government Anthropologist, a Scotsman whose forebears were in the India Civil Service. He fought in the Boer War, and worked in Nyasaland and Tanganyika before joining the G.C. administration as a Port Customs Officer in 1907. In 1909, he became Acting DC Yeji, along with other duties. A short, red-headed Scot with an excellent ability to speak local languages

but unable to adjust to the role of DC, meant he was never to become successful. However, he found his big role in analysing and explaining Ashanti Customs. He took a one year course in anthropology at Oxford in the pre-Malinowski era. His idea was to educate the Ashanti into their own culture, and encourage local tradition. Given the monikers "oboroni okomfo" (white witch doctor) and "Amoako" (red pepper), and in the NTs "Bo'adaan" (for his obsession for shrines). He first came to the northeast of the NTs in 1928, to prepare for his fieldwork, visiting the Tenzugu shrine asking for success in his endeavour to fly a plane solo from England; he also enquired of a diviner in Navrongo about his field-studies. Ref. Machin P154. It was the first solo flight to West Africa and he was nearly 48, and had been in the GC for 22 years. Stewart Simpson the Vet told Machin, "I had built up a mental picture of Rattray and imagined him to be a tall, tough-looking, perhaps bearded and altogether the sort of 'white hunter' who could eat a duiker for breakfast. I was indeed surprised when I did meet him for the first time. He was small, slim and soft-spoken, his facial features deeply tanned, I cannot remember the colour of his eyes, but I can only describe them as being keen and penetrating. Altogether he was small and slim, he was immensely tough and on my first hunting trip with him at Batisan, to the east of Tumu, he almost walked me off my feet." Ref. Machin p.173. Rattray went on to write up and publish two volumes, on his research, called "Tribes of the Ashanti Hinterland". In 18 months he had filled 34 notebooks and covered 15 tribal areas. He left just as Charles was arriving in the Gold Coast. So Fortes was not the first to study the Tallensi, but did so in greater detail using the then modern framework of Structural Functionalism.

Chapter 5 - 4th Tour 1935– 37

112. *Agricultural Adviser to the Secretary of State.* The post was created in 1929. Only four officers held the post before the Colonial Office closed in 1966. Sir Frank A. Stockdale was born in 1883, graduated from Cambridge, and was a one time Director of Agriculture, Ceylon. He held the posts of Deputy

Chairman Colonial Development Corporation, and Chairman of the governing body of the Imperial College of Tropical Agriculture, Trinidad. He had a great influence on all Colonial Agricultural Departments. He received the G.C.M.G and C.B.E. for his services, and died aged 66 in 1949. In January of that year he wrote the forward to Charles's "Agricultural Extension and Advisory Work, with special reference to The Colonies" Colonial bulletin 241, with the emphasis on increased productivity.

113. *Sir Arnold Hodson* (1881–1944) had wide experience in South Africa, Bechuanaland (Botswana), Somaliland, and Abyssinia (where he got the nickname "Ras"). He was Governor of the Falkland Islands, Sierra Leone and finally Gold Coast (1934–41). Not to be confused with Governor Sir Frederick Hodgson, of the Ashanti Golden Stool infamy.

114. *Harold Blair* (1902–85) was the son of Scottish Missionaries, born in India, and had a degree in theology and classics from Oxford. He took the Colonial Service course with special interest in languages, and arrived in Accra in 1928. He became a Dagbane Higher Standard speaker in 1930, and then later a Gonja speaker, and would go on to publish dictionaries and grammar books in both. His monikers were "Son of Dagomba", and finally "Red Yakabu", a reference to his shooting dangerous animals and for the pot. The Fortes stayed with them in Tamale in January 1934, where Harold was DC, and noted that Blair was "an excellent Dagomba speaker and a very charming English gentleman with a great facility for getting on with the natives." He had collected a good deal of data concerning the Dagomba and Konkomba languages and traditions, but all in "folklore" fashion. As a result he had a high reputation with his superiors like Major Jackson and the Acting CCNT, Duncan Johnstone. Harold was invalided out with amoebic dysentery in 1939. He became a teacher and then took Holy Orders when he left the Gold Coast, and eventually became Cannon of Truro Cathedral in Cornwall.

115. *Buligas/Bilise* were found in Yemen, Egypt and North Africa, so along the route for the Hajj. Honorary W.H. Amherst did research into bilise.

116. *Fortes and their fieldwork.* After just over a year in Tongo the

Fortes left for England in April 1935. On their return for a second tour on 14th May 1936 Fortes wrote to Miss Brackett, Secretary to International Institute of African Languages, London reporting,"Mr. Jones said he had found the papers extremely useful, and was passing them on to his DCs...Very encouraging to find we have been able to supply useful information and suggestions without, in any way, deviating from our scientific objective." In his very full report on the year's field-work, from April 1936 to April 1937 Meyer Fortes told of his tribulations on returning for the second tour. He and Sonia arrived at the beginning of the rainy season and immediately had problems with the rest house and outbuildings. Despite paying for their repair himself, the thatched roofs leaked badly. He even reflected a tent might have been better. The rains season was always the most difficult time of year. On top of that he had a serious attack of malaria, which was recurring throughout the rainy season. In mid October he had a particularly bad bout, his fourth, and DC Allan Kerr took control of getting him to Tamale Hospital. Sonia recalled how the police, runners and messengers were used to organise his departure by hammock to Shia, and then to Tamale. Ref. Fortes Papers CUL. Add. 8405/2/1937. While they were away in Tamale the motor road to Tongo was finally made up. For an insight on how the Colonial Officers generally saw the Fortes Jock Stewart remarked on 29th May, 1964 from retirement, after reading a letter sent from Fortes to Charles, "He did not like our old administration at all as the DCs were all inclined to turn up their noses at him. I did not like him at all at first but latterly, found him very interesting and not at all a bad chap." Ref. Personal Letters.

117. *Magical aspects of agriculture* had a very important role, with one Tendaana calling himself, "will drop manure". "He was comparing himself to a herd of cattle whose coming would bring fertility to the impoverished soil of his clan and so giving them more food and well-being." Ref. Names among the Tallensi p.349. Fortes. Farms were protected from thieves by different medicine. Some had stone boundaries, others had fowl feathers, horns, bits of old bed mat and so forth suspended on a stick. Ref. Natives p.86; Culture Contact as Dynamic Process pp.193 –6. Fortes; Clanship pp.168–190.

118. *Grass burning.* Trees had a special place in ancestor worship. Some held the spirit of the ancestors and needed libations at certain times, others needed to be "burnt". The Tendaana held the key role. Charles suspected that the tradition may have had a functional role in slave raiding times, when long grass hid raiders. It also created a short-lived fertility in the soil. There was a whole set of rules governing land tenure and ownership of trees etc, and all these traditions affected Agriculture in different ways. Grass was needed for thatching, animal bedding, and manure-making for mixed farming, and small trees could flourish for firewood etc. It also prevented soil erosion. The spirits were moved to Tengani groves. Due to its Protectorate status the Governor passed an Ordinance in 1933 for the prevention of grass burning, and in 1962 when Charles returned to the area for an assignment with FAO he was told it was still in force. He wrote to Jock Stewart, retired Director of Veterinary Services, that it was still being largely observed. Jock replied, "I am delighted to hear about the results of the fire control around Zuarungu which must have been particularly satisfactory personally as you originally initiated this excellent measure. I was also pleased to hear in your letter about the growth of trees, but is the Ghana Government preserving the forestry reserves down in the South?" Ref.12th January 1963 Personal Papers.

119. *"Agriculture in North Mamprusi."* Bulletin No 34. C.W. Lynn.

120. *Nangodi Mine.* On 25th July 1934, Sonia Fortes told of meeting a newly arrived man called Aplin. He claimed to have travelled widely in many countries in Africa, and had been a professional journalist making films and writing articles; he had also been a big game hunter for 18 years and bragged of having made the huge sum of £200 per month at it. Despite being a quiet man he showed a great contempt for the natives. "The thing that he said annoyed him most is that a native is treated so well here by whites. He would not allow a black man to be addressed as 'Mr'." Less than a month later Aplin had left, having passed through Tongo and spent the day with the Fortes. Sonia recalled that he was the type of man who one would not want to entertain at home, and less so in the bush. "A man who has

290

none of the minimum essentials for this country, and is only an embarrassment to any European who comes across him. I think he is a cadger and ready to live off the other fellow. He came here with less luggage than his native servant, I should think." He had simply a tin of milk, no clothes, a camp bed and a tin teapot with him. The mine seemed to attract such adventurers, but they did not stay long. Ref. Sonia Fortes Diary.

121. *Dr Denoon of Canada* lies in Tamale Cemetery, under a simple stone cross.

122. *Vegetables.* Most government officials had their own vegetable garden, and Charles used to tell a story of Meyer Fortes requesting vegetables. It went something like this:-

"Dear Lynn, Please send us some vegetables urgently, Regards, Fortes.
Being the dry season vegetables were few and of poor quality, but some were sent.
Dear Lynn, Is this the best you can do? Regards, Fortes.
Dear Fortes, Here are some seeds, you try? Regards, Lynn.
Dear Lynn, Seeds failed. Regret previous comments. Any vegetables? Regards, Fortes."

123. *The Agricola Club* is the Society of former Wye College Students.

124. *Plough in West Africa.* It was recognised that the native farmer had little capital for buying expensive imported ploughs, and there was a need for an affordable locally made one. Nigeria was ahead of the Gold Coast in designing a double-breasted ridging plough. It was also necessary to have different ploughs designed for the different types of soil. So after a lot of experimental work it was found that some could be made locally, fairly simply of wood: it could then be copied and manufactured by trained craftsmen. Before 1930 missionaries and other administrative officers had tried to introduce the plough into Northern Nigeria, but with little success. These had been imported and did not suit local conditions. A new wooden plough could be made locally for less than £1.10s 0d, and were quite satisfactory. They were tried out on the experimental farms in Northern Nigeria for some four years, and the local native farmers who tried them were satisfied. There were three drawbacks compared to the iron

plough; they dug deeper into the ground; they could shrink and crack in the changing climate, especially the dry season; they could be attacked by white ants. Ref. West African Agriculture p.67. Faulkner and Mackie.

125. *Sir Bryan Sharwood Smith.* Author of "But Always as Friends," worked in Nigeria 1921–57. My mother remembers him at the Secretariat in Lagos, when she was private Secretary to the Governor . She recalls that after a dinner at Government House in 1936 the Governor withdrew, and Sir Bryan and some others were involved in a feather cushion fight leaving the room full of white feathers!

126. *Imperial Airmail Service* was by the double- winged, propeller-driven De Havilland Dragonfly.

127. *Polo* was also popular with the local Hausa ruling class, and ponies were locally bred and trained.

128. *James Richard Mackie* CMG Nigerian Agricultural Service 1921–45; Director 1936–45. Co-author of "West African Agriculture" with Faulkner.

129. *Taxation.* Five months into his second tour in August 1936, Fortes was in Zuarungu paying a visit to the Kunaba: also present were the important Chiefs the Tongraana, Sieraana and Quusnaaba. He took the opportunity to ask again about what they thought of tax collection, which had begun in Biuk. Fortes first asked Kuna what he thought generally about taxation. " He said, noncommittally at first, that it is a good thing; anyhow, no one can refuse the white man's order. But he did not know why it was being introduced. When I pressed him he said it was to build roads, wells and good houses – all of which he said was very good, so that they could have proper roads free of snakes etc. good water and good strong houses." He then turned to the other three chiefs, and asked what they thought "... they said – Sir we are dependent on what you say. Nantuama was scornful, saying they had been told several times by the DC what the tax is for but could not understand it."

The imposition of tax at this time created insecurity and suspicion, and to add to the problem locusts were ravaging the countryside once again. On 1st October 1936, Fortes wrote,

"Tax collecting is approaching and all are preparing for it. People hoeing new bush farms this year, say it is because of the tax. Many go to market with produce saying – how true one can't tell – they want to sell in order to collect tax money. It is a frequent topic of conversation. At a funeral compound the other day… A group of Seug men were sitting, and one or two others. When I came up the conversation turned to the tax. There was a general outcry (we can't afford it, they will have to take me to prison) some said, this is only a beginning. Next year it will be 2/- and next 5/-. Kuyksydap of Zubiug was present and began aggressively to argue with me. 'What debt have we contracted,' he kept asking, 'for you to be taking our money thus?' Then he began delivering himself to the audiences, adulating examples from his travels to the Coast and Ashanti. In places like Kumasi, Sekondi, Accra people swarm (like sand); they have clothes and food and everything you want. Will they too pay 1/- each? And what about Fra-fra folk who live in Kumasi and elsewhere? And what about a stranger who just happens to come and live with you here for a while? It is unfair. In what way are we in debt of the white man? In any case (there is no money here.) The white man has plenty of money – what about his (gold mines) full of gold? What about his Banka (Banks), full of money. The white man should find money for us, not we for him. All these propositions were argued about. Some of the others were in favour of his arguments; others objected and most just listened and by further questions satisfied their curiosity. But some tried to explain what, if any, was the importance and the value to themselves of the tax. The tax collectors have reached Baari. The Chief (of Tongo) was very offended. He told me that Kuna's elders, who are collecting there, had never informed the Tongraana about his intention to collect at Baari. Of course he blames the DC. If the DC gave his orders that the Tongraana must be informed, he would be informed." Ref. Fortes Diary. CUL.

130. *Locusts*. These were a huge problem for agriculture, with no easy eradication solution. Fortes recalls in his Diary on 28th October 1936, "Locusts have been ravaging the whole countryside this last week. They have done tremendous destruction in the Bongo and Nangodi districts, where the harvest is later than

here. I saw them, a great whirring cloud, rising from the paths and fields at Zuarungu on 26th October. Yesterday they reached Yamlaga and destroyed a great many bush farms. Last night they reached Mo'aduur destroying bush farms there. Thus talk has been solely of locusts these last two or three days. Messengers have been coming in to the Chief to report, and he has been sending to tell the DC at Zuarungu. Today Emer from Wakii passed to report the destruction at Mo'aduur of nine Wakyii bush farms. He was very rueful saying 'Does it not warrant killing one's self with a poisoned arrow?' And indeed it is said that two or three men in Nabroog killed themselves with arrows when they saw their crops vanish thus. As to what to do about locusts, the reply was, "Only you whites can solve the problem for us." Generally speaking, there is no anticipatory anxiety and terror about them, as there is about drought; there is grief in reaction, among those who suffer, not anxiety on anticipation. (In drought there is famine for all; locusts' selective mischance.) Locusts, said Yikpemdaan, don't come every year. Sometimes they may not come for a great many years; and sometimes they come several years in succession. Some 35 years ago locusts did so much damage that there was a famine and people caught one another to sell." Ref. Fortes Diary CUL.

131. *Stewart Simpson*, the Vet, nicknamed "The Simp". He and Charles lunched with Fortes in August 1934, where he vented his frustration at the lack of results for all their efforts, recorded in Fortes' Diary. He was friendly with Rattray and went hunting with him. Both were redheads, and Scots.

132. *Car maintenance*. It was essential to be able to repair and service your own car. Kumasi was the nearest main garage, but the local truck drivers were excellent and helpful mechanics and always ready to stop and help. The local blacksmith was also competent for all welding and metal jobs.

133. *Loss of dog*. Rattray noted, "I need not try to explain to anyone who has ever lived much alone in Africa what the companionship of a dog is to his master, and what Adinaya's loss therefore meant to me." His red pie bitch was taken by a leopard from under his bed in the rest house in Tumu. She was so badly mauled that she died in his arms. Ref. Machin p.70.

134. *Cattle ranching* caused a big row at the Political Conference in December 1936. However, in a personal letter written in 1962, Jock Stewart wrote to Charles "I have always been against ranching in the Gold Coast ever since, shortly before you came to the country U.A.C. prompted by the then director, General Grey, wanted to take over a large area of land in Builsa, and of all places, to start ranching. Jacky, who had just become C.C.N.T. was sympathetic, so was the Governor, Shenton Thomas and I had just been appointed DVS. I was dead against the scheme and did not yield an inch and Percy Whittall, was PC Northern Province, and the DCs were also against it so it fell through.... Even when there are large areas of land more or less unoccupied I am against ranching. I have seen its ravages in too many parts of the world, not least in my own Scots Highlands where the people were evicted and sheep ranches established which has converted the Highland hills into a man made desert, the sheep have drained the land of its fertility, which was built up in the old days when the Highlands was thickly populated, when the peasantry kept cattle and goats and went in for mixed farming." Ref. Personal letter from Jock Stewart to Charles dated September 1962. Report on Gold Coast Livestock. J.L Stewart 1937 CO96/740/11, PRO.

135. *'Kibi'*, Sir William Jones, a short dynamic hardworking Welshman, was at the centre of great change with implementation of indirect rule. Political Conferences bound all the Departments together, and social reliance, friendship and discussion did the rest. An interesting true story is told by Jock Stewart, "By the way Ebenezer Adam is Regional Commissioner of the Northern Region, in other words C.C.N.T. That is the ex Tamale school boy who wrote an unfavourable essay on Native Administration, when Kibi had the top class all writing this essay, and refused to change his charges of corruption etc. against the chiefs. Kibi had him sacked from the Tamale School and exiled to Kumasi to be placed under the guardianship of the Sarikin Zongo. However, I wrote to Pansey Townsend to see if he could help the lad. As you know, the SHO was feared more in the Zongo than the CCA, so the Sarikin Zongo sent Adam to a good school and he eventually became a teacher, joined Nkrumah's party and now is one of

Kibi's successors. I occasionally get a letter from Ebenezer Adam who is a very decent chap. I may say that whenever I write to Kibi, I do not mention E. Adam!" Personal Letter to Charles dated 9th September 1963.

136. *Visit of the CC and Nana Sir Ofori Atta* to Zuarungu. Fortes had gone to Zuarungu to confer with Kerr about administration matters, when he heard the news that Chief Commissioner Jones was giving a guided tour of the region to Sir Ofori Atta and the Omanhene of Winneba. Jones had received his monika "Kibi" from working on the introduction of Native Treasuries etc. in Sir Ofori Atta state. It is ironic that the murder that followed the latter's death in August 1943 would be one of the death knell to indirect rule itself. "Yesterday a very distinguished party passed through Zuarungu en route to Bawku under the guidance of Jones; Sir Ofori Atta and the Omanhene of Winneba. Jones is showing round his native administration here – an object lesson, I suppose. But they wanted to come back and visit the Tong Hills. As Kerr said, probably to see the Fetishes....." Fortes again on 23rd February 1937, "The great cavalcade of costal royalties arrived at about 9 am escorted by Kerr, and with a large lorry full of hangers on. They wanted to go up the hill, to the fetishes as Kerr suspected. How significant that two 'enlightened' modern chiefs - members of Leg.Co.- should come all this way, ostensibly to examine the workings of the Native Administration - and be prepared to devote a whole valuable morning to paying respects to the famous Tong Fetish. They went off with a flourish in their cars to the bottom of the hill; returned at 1, all (except Kerr) spotted and marked with the red mud - and apparently quite pleased to be thus differentiated and quite unembarrassed not only they, but all their swarm of followers. The Tongraana waiting for them with drums etc. and with grand magnificence presented them with a sheep each... They then went on to the Golibdaana, who was awaiting them all dressed up in his gala garb. The DC apparently was very off hand. The pilgrims insisted on going up to the boyar, but Kerr was tired and preferred to stay down below and wait.

At the Golibdana's compound, Kerr refused to climb up to the boyar, but the pilgrims insisted on going. Kerr asked him to be quick and sat with Yikpemdaan waiting for their return. So they went up to that boyare too; and when they returned, they gave 10/- to the Golibdana, who asked if this was for himself or for the boyar. They said, he must decide what to do with it. Then he produced some fowls to present to them, but Kerr absolutely refused to allow gifts. So they went. On the path, Nyaayzum, who had also received a dash from the pilgrims, intercepted them with a sheep, 2 fowls, 4 yams and eggs. He insisted that they take it, saying that as they had come to Kpataar naabis bagre and had entered it, so it would be improper to refuse the gifts. In short, Naabdia indicated that the Golibdana got very much the worst part of the bargain and Kerr showed great wisdom." Some years later Allan Kerr would be involved in the Kibi Murder Case, when a headman was ritually killed to go with Sir Ofori Atta to the other world.

137. *Lynn and agriculture.* Fortes reflected on the uphill battle Charles faced, and how he found it so difficult to accept the slow change in outlook of the people when food shortages resulted. The social obligations of ancestral worship were of the greatest importance to Fra-fra life, and were resistant to change. The great care taken of the tobacco fields showed what could be achieved. On 18th February 1935, returning from Tamale, Fortes concludes, "A dreadfully tedious journey home, through thick harmattan haze. Passing a never ending stream of men going south carrying chickens on head – many pedestrians, Lorries crowded; others going north again, an endless procession, Lorries full too. Such is trade, culture contact – the factor which specialist Agricultural Officers ignore when making plans. Why doesn't the Government organise and develop this aspect of NT economics. All along the route, at every Dagomba village, women sit under trees, selling food for the passing fowl dealers. Some of these quite young boys, carrying half a dozen chickens, no more.... Between Bolga and Winkogo an enormous riverside area laid out in neat tobacco fields. Probably this is what enrages Lynn who can't understand why they devote so much immense care to tobacco, and neglect food crops. Perhaps tobacco needs it!! But Lynn says they are too lazy, working only 150 days per year, because farming is a custom with them, not a business! Does not understand that in a society which has no social division of labour, with farmers as specialist body, it can't be a business – as Yikpemdaan's cap

making eg. is, some peoples' cattle dealing, salt dealing. Hausa Trade." Ref.Fortes Diary

138. *Fortes left the Gold Coast* in April 1937, to write up all his analysis of the Tallensi. His old friend the Tongraana died in 1941.

139. *Gordon Cowan* was five years older than Charles. He was a keen sportsman, a Scot, who lived in Echelfechen. He was one of the first to graduate from the Imperial College of Tropical Agriculture, Trinidad, and was mostly stationed in Tamale. He was due to publish his Bulletin on South Mamprusi, centred on Tamale. There were doctors at Tamale Hospital and dispensers, but no permanent nurses, and Princess Marie Louise laid the foundation stone in 1925. The granite surround to his grave came from Scotland, and still remains in Tamale graveyard, but it is unmarked as the brass memorial plack has disappeared. His nephew Alec Murray was to be Charles's Deputy Director of Agriculture in Nassau, Bahamas in 1966. Colonial circles were intertwined.

Chapter 6 - 5th Tour 1937–39

140. *Taboo against work* on the farm for fear of father's wrath was so strong. A similar story of ancestors' wrath from Fortes' Diary recalls this situation on 22nd February 1935, " At first the old Tendaana, a blind old man, said nothing. But soon he came out boldly, with the attractive courage of one who believes firmly in what he says, has the courage of his convictions..... The Tendaana explained more fully. Three years ago he saw the naara (early millet) sown. Now he is blind. He consulted the soothsayer, and the gaga said that it was because they had thrown away their ancestors. He had eyes to see and find his way to Bolga, to a strange chief, a Nankani. His ancestors would deprive him of his sight, so that he could not see to find his way there any more. Not only that, but every time somebody dies and people are dying in great numbers there, it comes out. We are dying because we have thrown away our ancestors." Ref. Fortes Diary. Clanship p.125.

141. *Anasuri* went to work at the mine. Sonia Fortes recalls on 25th July 1934, "It was rather amusing to find all the rejects from our and Lynn's household in his retinue. Alassan, our erstwhile cook at £2.10/- per month – their head boy at 30/-. Lynn's cook, sacked for too much pito, their cook, steward boy and wash boy for a price, I should think much less than Lynn gave him. Isaka, who worked for us for about 4–5 days at the rate of £2.10 a month, at 4d per day! Tete, who very nearly became our cook at £2.10 a month, for 25/- p.m." Ref. Sonia Fortes Diary.

142. *Hyenas* did take dogs when they could. Allan Kerr was sleeping out with his dog in Zuarungu, when a hyena stole up and tried to take it from next to his bed. He awoke, grabbed his gun and gave chase down the tree-lined avenue in his pyjamas. In the affray he broke the butt of his gun! A story retold many times.

143. *Agricultural Development* finally became a presence in the Zuarungu District. Tamale's research station had not produced such tangible results. Fortes, writing to Malinowski on 2nd September 1934, shows the activity at the start of the farming season, "With the coming rains, the energies of the community were fully absorbed in agriculture, the staple industry of the country. Men, women and children have been fully employed in the various processes of hoeing, sowing and weeding. Starting from direct contact with family groups actually at work in the fields we soon got into problems of household economy, organisation of family work, kinship rights and duties, and land tenure. We had previously attempted to start day to day records of a limited number of families with which we had established cordial relations. But we made little progress until this phase of economic life gave us a firm basis to build on. It was in such activities that the unity of the joint family manifested itself to the eye. Every field has its correlated social groups, owning it, working it, and preserving it for posterity…" Ref. Letter in Fortes papers Add. 8405/1/45/12 (iii) CUL.

144. *European women in the Gold Coast*. The Governor in 1911 made a general order decree that no European ladies should stay in the Northern Territories, citing lack of creature comforts. In the 1930s women went home for the difficult rainy season. The biggest change came with Sir Alan Burns being appointed

Governor during the war. He personally understood the problems of this situation, as it was his third tour in Nigeria before he asked permission to take his wife out with him. He recalled the problems as a shortage of housing(usually in the south), where some junior wives had to be sent home: In outstations married officers with wives often got preference, making some Senior Officers have inferior quarters; older officers objected saying it affected their husbands' work, so he did not travel as much – not wanting to take them or leave them behind; "Occasionally the real objection of the hardened 'old Coaster' was to white women as such. He regarded them as intruders in to what had been essentially a bachelor's paradise, where a man could dress as he pleased, drink as much as he liked, and be easy in his morals without causing scandal." He reflected that perhaps, because they did not have to consider the comfort and prejudices of the white women, he was on better terms with the Africans; so he visited more villages and entertained more Africans. A wife's illness, quarrels with other officers and/ or their wives, living in small communities that worked together could be disastrous. Ordinary difficulties in married life could be magnified, with boredom if wives found nothing to do. But men lived better and happier lives with their wives. Servants could resent a wife when previously they held total control. Wives could demand cleanliness and a more efficient routine, and be upset by carelessness. Ref. The Colonial Civil Servant, Sir Alan Burns.

For numbers of women on the Gold Coast see 1921 and 1931. Ref. Census P255. No officers going out for the first time had wives with them. Indeed, it was only after two tours that most officers married. Times at home were very short for finding and making relationships, and when children were born it was more problematic as they often had to be left at boarding school, fostered or the mother had to remain at home with them. When Sir Alan Burns became Governor of the Gold Coast in June 1942 he let it be known that he had no objection to children joining their parents. This would allow Sylvia (Lynn) and Michael (Syme) to live with their parents in 1945. While they were liable to get malaria, prickly heat, and insect bites, as well as suffer from the hot and often damp climate, the decision should be left to the parents. Nothing should be done, however, to encourage them. After all, white French children lived in the neighbouring French Colonies.

145. *Cattle wealth* and its effect on bride price. Rinderpest was being brought under control by 1930 so cattle became healthier, but large numbers were used for bride price. A successful man was judged by the number of his wives. Those poor in cattle had to give an IOU, and wives could be removed if debts were not honoured. European-style money was coming more into use, but cows were seen as very important to the economy, and essential for sacrifice and inheritance. Now their potential use as draft animals gave them an added value. Cattle were traditionally driven down from French Territory for sale in the Kumasi market.

   Fortes records in his Diary on Tuesday 29th February 1934, "In the course of such breathing space Anaho slipped an item of exceptionally interesting information. He told me that his brother Chief SOMBORA had only 4–5 cattle now. The old chief had left 30–40 head of cattle when he died, and so many sheep and goats – he said 1000 – that even if one got lost he didn't care about it. Sombora however had used up most of the small stock to get wives – he has 7 now and had only 2 when his father died. The cattle he had used to buy his Chiefship. He had paid the men who had 'begged the Chiefship for him' from the DC – namely Bassana, 6 cows and the Tongo Na 4 cows. Thus is the new Native Administration being founded upon ancient and honourable traditions." Ref. Fortes Diary.

146. *The Native Administration School* in Zuarungu was created from taxation and replaced the old smaller schools. Agricultural development was linked to all schools. A vegetable garden was also found in all the schools. This was once again as a direct result of indirect rule, and the formation of the paid Native Administration.

147. *Palaver* was a Portuguese word for "talk". Commonly used for group exchange meetings, sometimes including elaborate greetings, drumming and dancing, brought into the language in slave trading days.

148. *Bassana Moshi* – also known as Grunshi. This name betrayed his dual roots in the community; he claimed his village was near Ouagadougou, and returned on occasions. Bassana was

what is now called a "child soldier." Ref. Cultural Affinities of the Tallensi Clanship P4–6.Fortes recalls in his Diary on 31st January 1934, "We got hold of Bassana Moshi, ex-Company Sergeant now settled here as the DC's factotum – supervision is everything. Thus he is in charge of the rebuilding operations on the rest house, if any. Old Bassana, probably in his 50s, is a shrewd old fellow. He interpreted for us on our visit to the Chief of Zuarungu, and acted the guide, with as much self-possession as any European guide to one of the more frequented show places of Europe. He interprets, in his semi pidgin, with the skill and self-possession of an old hand. But I am quite sure his versions are summaries of salient points or of details. Alas he has been Court interpreter here and has acquired a legal fondness with words. Bassana has evidently done well for himself. His son is a PWD carpenter at Tamale, another is I believe at a school, either at Bolga or at Gambaga. He claims to have a large retinue of wives – Nankani, Nabdam & Tallensi – probably fairly well to do." Ref. Fortes Diary.

On 29th February 1934, Sonia Fortes records, "Bassana came on Thursday morning to tell us why he was not present at Tongo the day we came there. He said that the people don't want to work. When he came there one morning he found not a soul at the place (Tongo Rest House). He got onto his horse and went round gathering the Kambonaabas. 'I flogged them to make them collect the people to work.' Flogging seems the method widely adopted here.....So Bassana is quite well connected. He is apparently the local aristocrat. Every one calls him 'Salmeja' – Sergeant-Major. One of his sons was a policeman at Accra and has been transferred to Z; another of his sons is a carpenter, trained on the Coast. He regards himself as the DC's Deputy and is apparently quite high handed with the natives. According to Isaka he is not a proper Moshi. He calls him a Keparsi. His father, he says was a Grunshi and his mother a Moshi; he came from the French Territory. 'There be too plenty humbug' from the white man. They catch the small Pickens for soldiers and force the people to do work without pay. There are Moshis, he says everywhere, here and on the Coast. They ran away from the French."...Talking of the eradication of Tenzungu, the Hill Shrines, Sonia again recalls on 26th December 1934, "Yet they have been afraid all those years to sacrifice on the spot surreptitiously – in case they said, the fowl feathers would betray them. Bassana – who claims to be Mohammedan – was actually the man who supervised the smashing down. He tells the tale with great gusto and in perfectly cold blood – No wonder he is held in holy awe by the Tenzugdem." Ref. Fortes Diary also for more details see Ref. The Black Volta at the beginning of Colonial Rule p.228, Goody, for ribbon replacement for Victorian Jubilee Medal, Ref Letters p.106.

149. *Rest Houses* differed greatly. Yinduuri is described thus by Fortes, in his Diaries, in December 1934. "Yinduuri is indeed proper bush. The wild country in the harsh unshadowed light of noon, looking garishly desolate; desolation of the spirit rather than of appearance – desolation of stark boulders. Tall dry grass, lonely trees like random scarecrows on an unploughed field, and desolation not of the contourless desert, but of a neglected wilderness. Thus on one side, on the other, bare hills the stony surface of which is cultivated to the very top – also desolate now, with a desert like nakedness. And the rest house is as uncomfortable and as dreary a place as its surroundings, the rooms all awry; their floors sloping like the deck of a pitching ship. The top of the hill is a broad and well wooded plateau like a flat saucer, rimmed with low ridges. It has the richest collection of dawa dawa trees I have seen hereabouts." Ref. Fortes Diary.

The Rest House at Tumu was also rumoured to be haunted – by the Ghost of Ferguson, a lion hunter. Captain Swire was buried under a mango tree in the compound. In the 1930s that area had many lions, leopards and elephants, and they all came into the Rest House compound, as Rattray recorded.

150. *Detokko* This is the complicated history of one of the chiefs with whom Charles worked, but it shows the great differences between the chiefs and their ability to make changes. Fortes notes in his Diary on 27th March 1937, "The Detokko naam (authority to create a Mamprusi chief) according to the chief is a very unsettled affair, as Detokko has become a settlement during the last 50–60 years. Men of 50–60 yrs knew, as small children, the original settlers at Karbok. The Chief's own father Doo was

the first of his line to come and live here, at the present house, while his junior brother by same mother Mo'naab, went some distance off, about a mile away, to build. These two were the sons of Too and their line is Toobindem…This is a fine example of small groups moving out to find fresh 'wild' land to create new settlement." Ref. Fortes Diary. "Needless to say none of the prohibitions, taboos surrounding the Tongraana hold here. The chief farms, as he likes, walks on the earth (ie Unshod) etc. He is extremely unofficial and an undistant chief, joking with his fellows – one amused by a joke grabbed his shoulder. The chief has his own earth shrine, the Karbok Kolog which runs near his house. This is not connected with the naam, but is his own inherited shrine from his father when he came to build here, and established the shrine."

151. *Fowl taboo.* "It is recognised by all Tallensi that these taboos are on a par with the fowl taboo of the Namoos, though they have, in many respects, a more complex meaning. Among both groups of clans, no doubt because totemic taboos are so clear-cut and categorical, it is considered unthinkable for an adult in full command of his faculties to violate a totemic taboo deliberately…since all Tale women avoid eating the domestic fowl from the beginning of pubescence onwards. It is one of several food taboos observed by women with the utmost strictness, not for ritual reasons, but purely as a matter of feminine propriety." Ref. Clanship p.125; Natives p.39.

152. *Conversion to Christianity.* Sonia Fortes Diary on 9th February 1937 recalls, "Nabila was telling us how in 1934 while his master was on leave and he was visiting his family he made his first acquaintance with Christianity. A catechist visits Zuarungu twice weekly and any one may attend and learn. He went because he was told that if you are a Christian 'you go to a good place' after death.... Yet the price he is asked to pay for it he considers too high: to have only one wife and not ever again give water to his father. When he was told that he stopped attending the meetings, and has resigned himself to hell. How can I chop and deny my father water, he said. That he can never agree to do, apart from the fear that his father's wrath might come down too heavily upon him..... But even Ayeltiga is irked by the

monogamic restrictions and says that he wants another wife but that the Fathers wouldn't agree to that." The people had no set beliefs about life after death, and Christian ideas brought conflict: "unless they are ready to give up what constitutes to the natives the core of their existence, Kaab bagere (making sacrifices) and wives which are the two essentials from which the Fathers wish to convert them. It is interesting how much these two factors do mean to the people. In the case of Ayeltiga conversion meant nothing since he was still young and had neither wife nor any occasion to Kaab (make sacrifices) .... Conflicts between the pagan survivors and the Fathers are not uncommon. "Ref. Sonia Fortes Diary. Ayeltiga was to become Chief of Zuarungu, some years later.

153. *Fishing festivals* and ceremonies. To Malinowski, 12th June 1934, "In the course of enquiring into the seasonal economic activities of the dry season, we found that we had arrived just in time to take part in several communal fishing expeditions. We adopted the customary procedure of getting a preliminary account from an informant, and of checking our observations subsequently with the help of an informant. This was the first opportunity we had to apply the plans we had drawn up as a whole. It gave us, as it were, a concentration, in one event, of every aspect of economic activity, very clearly defined.

Thus the ownership of the fishing pools was vested in a particular individual in virtue of succession to a particular chieftainship. This ownership was obvious and emphasised in the way the fishing expedition was organised. We anticipated, searched and found a magical sanction for ownership of pools. We had been told that anyone who wished could come; but then we analysed the 'personnel' of the great concourse of people, we found that it was strictly determined, on one hand, by a political cleavage which divides the Tallensi into two major groups, and on the other, by personal kinship bonds. Following up these clues we were able to establish a preliminary outline of the political structure of the people...It appears that the political line of cleavage corresponds to a 'totemic' line of cleavage... The ancestor cult is also the principal magical institution. We first saw it in action on the fishing expedition where we found that the owners of the pools were responsible for the protective magic which guaranteed no accidents and good catch. Later when the rains failed to come the chief appeared in the role of master-magician and rainmaker, in virtue of owning the supreme ancestor shrine." Ref. Letter to Malinowski Add. 8405/1/45/12(i) CUL. For further information see Ref. Communal fishing and fishing magic in the NTs of the Gold Coast, Fortes.

154. *Markets.* Fortes to Malinowski, 12th June 1934, "It has been a more difficult task to deal with an institution like the market; its function as an 'exchange' centre in the economic system was clear from the outset. But as soon as we began to investigate the relations between the production of market goods and their exchange in the market, we came upon an intricate network of social relations underlying the purely economic transactions. But the economic function of the market is only one. It is a meeting place for all and sundry, where gossip and news are exchanged, love affairs initiated, and communal enterprise canvassed. We have made various attempts to investigate this side of life. Our most successful device has been to station ourselves at a bottle neck on market days and make a census of everybody going to market, finding out what was his or her purpose, and later checking on their return to what extent they had carried out their purpose. This led us to another discovery; that the market has its regular frequenters who never miss a market day, and as we suspect, play a considerable part in disseminating news. We are continuing this market study and expect to get useful material from it." Ref. Add. 8405/1/45/12(i) CUL.

155. *Gologo dances.* Charles often recalled that in 1936 he arrived, with Allan Kerr, to attend the Gologo at the height of frenzied dancing and chanting. Meyer Fortes had been attending the complete cycle and was in the "thick" of the ceremony. The whole ceremony seemed to quieten down on their arrival. Fortes later confessed to Charles that he was pleased to see them both, as it was all becoming a little too overpowering. Fortes wrote to Evans-Pritchard later that year about the Gingaung Festival, 2nd September 1934, and expressed this again. "One of the practical limitations to conscientious field work we find is sheer physical inability to go the pace. I had

a series of funerals recently that wore me out so that I went down with fever, fortunately not a severe go, but bad enough to knock me out for a week. Just today a great annual festival has begun the main popular aspect of which is a dance every night, and all night, for the next six weeks. At the same time, the neighbouring settlement is celebrating a similar annual feast, with different dancing and in its own fashion. How we shall sustain the fatigue of one of these affairs only I don't know yet." Ref. Letter to Evans-Pritchard. Add. 8405/1/17/5(i).

156. *Dancing horses.* Marjorie's horse was doing the common West African trick of dancing on its hind legs. A vicious bit usually made the horse rear. Marjorie had just learnt to ride, and was fearful of horses. She hung on saying, "Dear God, please don't let me fall off." After the gallop back to the village, over yam mounds, she could not sit down for over a week and had to sleep on her stomach. She preferred the bicycle for some time after!

157. *Changing Administration.* A new local Court and Native Administration, funded by taxation, were set up; new schools, better human and animal health, and expanding improvements in agriculture generally developed. This was throughout the North, with new research stations planned for all areas, and new staff employed. "Kibi" Jones had orchestrated the change: his version of indirect rule. What was seen as lack of interest from Accra was superseded by strong leadership from the Chief Commissioner in Tamale.

Even after he left, Fortes was still in demand for his knowledge, and received in return any help and data he needed. The North was not represented in the legislature until 1946 when the Chief Commissioner became an ex-officio member of the Legislative Council. However, the Governor alone continued to legislate for the NTs until 1951, when all legislatures for the country became combined, and the first 19 members for the North were elected to the Assembly from electoral colleges. In 1954, it increased to 26 members when rural electoral districts were established in the North and a system of direct election was introduced.

158. *Drought and prayers for rain.* Fortes Diary recalls many references

to rain in 1934. Sunday, 6th May 1934, "Visited him (the Chief of Tongo) today and when I referred to rain, he said can't I do something about it? I said I had no power. He said I did. I could write and it would get to God. Pumaan and others more naive expressed the same sentiments. I am next to God. I, the Chief and the Tendaana own all. Won't I let rain come, so that food may abound. Yet with this dire and probably recurring necessity pressing upon them, there is no elaborate system of rain magic to 'control nature' or to give confidence to man. The universal technique is the making of sacrifices to the shrines. On the market it is becoming monotonous to receive in reply, so frequently, the answer I am buying a chicken or a goat – to make a sacrifice."

17th May, "As there was no other business I sat down and questioned the chief – leading on from rain to famine to history – a context, again, in which the chief's emphasis on his own significant value for the country and for the well being of the people emerged in what seemed an honest statement of historical facts. Every one of these palavers reveals how concretely and dynamically tradition – as known, invoked, cited and wielded in argument – enters as a steering, directing and decisive agency in transforming a situation of deadlock. The Chief in particular constantly cites the past – ancestors, traditions, and recent history, as fact, never as legend or mere tradition – in argument." Ref. Fortes Diary.

159. *Rabies* The word rabies comes from the Latin for madness. Louis Pasteur found the antidote. The injections were into the stomach, and very painful, but death and its association with a water phobia was truly ghastly.

160. *Winkogo Rest House.* "After Kpataar, Winkogo was very disappointing. The rest house is old and tumbled down, the landscape flat and uninteresting and it lies directly on the main road." Ref. Sonia Fortes Diary, October 1936. In a private paper to Ellison, Fortes explains the case of Winkogo chieftainship. Winkogo is seen as a "son" of Tongo. The area was bush when the original settler arrived, from Tongo. It is a single homogenous group. The chieftainship and tendaanaship are held by the same clan – and family. The land is held by original settlement. Naam for the Chief of Winkogo is given by the

Tongrana. The Bolegraana would like to choose the next Chief, as the final funeral ceremony has not taken place; and Fortes remarks it would "leave a lot of bitterness" and would be seen as an injustice and discrimination against Tongo. It is perhaps because there is no Chief that the rest house is not in fit order, as it should have been.

161. *Trees and Tengani groves.* "And it is worth adding that particular artefacts, stones, trees, and other material objects can like animals serve as the vehicle of living ancestral presence. A tree may be such an ancestor's shrine, or rather, metamorphosis. Its fruits may, then, not be eaten by his descendants. Particular portions and items of the natural environment and the material culture are thus incorporated into the sphere of a person's or lineage's moral and ritual commitments thought the ancestor cult and quasi-totemistic observances" Ref. Morality p.132 Fortes. "The ebony tree is believed to be a dangerous or evil tree, liable to be magically animated and then to injure or even kill people." Ref. Morality p.171 Fortes.

Chapter 7 - 6th Tour 1939–41.

162. *Scottish Colonial Officers.* There were many Scots in all departments of the Gold Coast. The Irish, especially the Catholics, tended to emigrate to America, while the Scots turned to the British Empire for jobs and advancement.

163. *Women in the War.* Women in wartime were not encouraged to come out, as it was difficult to get home in any safety. Marjorie's return in the middle of the war was to serve in the Women's Royal Voluntary Service,WRVS, and only on the grounds that she was needed by the Administration.

164. *Roland Smith* grew up in Northern Ireland. He would take over Charles's position in 1947. Latterly he served in Nyasaland and Borneo. Ref. Unpublished Report File 463 1948 Report on Agriculture in Kusasi. (Manga) R.Smith.

165. *Three agricultural stations.* Because of recurring famines, Zuarungu was the first station to be developed. After the survey

and experimental plots, research stations were set up at Manga in Bawku and Tono in Navrongo. Later, a station was developed in Babile near Lawra, in the northwest.

166. *Army recruits* from the NT's. In the Second World War the Gold Coast Regiment fought mainly in East Africa and Togoland, but in the Second World War they were sent all over the world. Ref. The Gold Coast Regiment in the East African Campaign, Sir Hugh Clifford.

167. *Gologo dances.* "There is no question that these festivals are the most important political institutions of the Tallensi. Every day I receive confirmation of the fact that Administrative arrangements which run counter to the alignment of clan-settlements in these festivals meet with resentment from the people. These festivals also have great theoretical interest." Ref. Fortes Report on fieldwork April 1936 to April 1937 Add. Mss 8405 CUL.Morality pp.54–65 and pp.46, 48, 100. "The continuity of the social life, from year to year and generation to generation is built on a process of fission and re-integration which had already become apparent to me in 1934" Ref. Fortes Report on fieldwork April 1936 to April 1937. Ref. Add. Mss 8405 CUL.

168. *The tsetse fly* is a blood sucking fly occurring across Tropical Africa bringing trypanosomiasis, (Sleeping sickness) a life-threatening disease for humans and horses, carried by game. Left in the soil to mature it emerges 22–60 days later, dependent on conditions. Prefers wooded shady areas near water, making bathing and water collection points very vulnerable. It was important to keep the area open once cleared, and farming was the best solution. The people needed to understand the life cycle of the fly, like that of the mosquito. Ref. Big Game and the N.T. farmer. 1946 in Farm and Forest 7. P 21–7. Morris. River blindness, found in similar areas, is caused by the parasitic worm, Onchocerca volvulus , perhaps brought to West Africa by the returning African slaves from Central America. Cattle and game found infected mainly in riverine areas, often transmits it as part of its life cycle. It is a disaster for the family when a person goes blind, as he becomes a burden. The loss of agricultural land near rivers leads to it being deserted, and a resource is lost, but still women usually have to go there to

collect water. Ref. River blindness in Nangodi, Geographical Review (1966) pp.399–416 Hunter.

169. *Elephants.* Wooded areas near water were the favourite areas for elephants. Chiefs often asked for rogue elephants to be destroyed as they did not have the fire power to do it themselves.

170. *This war* – The Second World War was white man's palaver, yet the people of the Northern Territories were involved, and saw Europeans doing unspeakable things to each other. It undoubtedly undermined white authority and domination. Jock Stewart wrote in reply to a letter from Charles, "Even at the very start of the last war, I was dead against bringing the African in any numbers, if at all, into a quarrel which was entirely that of Europeans. It was this bringing them into the war that set the ball of African nationalism rolling and now it just cannot be stopped. The same thing plus the fall of France and the ignominious defeat of the suzerain power made it quite impossible for the French to continue to hold their colonies." Ref. Personal letter to Charles. 27th June 1962.

171. *Car ferries.* Accessibility all year round was part of the great improvement in the general development of the NT's by the end of the 1930s. Fortes had pointed out the importance of produce and fowl trade with the South; ferries now made it easier to travel, trade and to collect data. Tolls and movement of produce etc. collected in Bolgatanga in Watherston's time now more easily and reliably taken at ferry crossings. Ref. The NT's of the G.C. p. 363 A.E.G.Watherston.

172. *Chief of Bongo, Anane Fra-Fra* again! Indirect rule took away his traditional privileges of tribute. His extortion and corruption led him order the murder of a young man, whom he accused of stealing his cattle. He was imprisoned for a year and destooled by Acting Chief Commissioner DC Guthrie Hall in 1942, aged 80. Ref. The 1916 Bongo "Riots" R.G. Thomas.

173. *Babatu and Samory and Autochthonous Societies.* Pre- British Rule, some indigenous people had solely a bow and arrow culture, for warfare and hunting. This made them vulnerable to societies with centralised politics, superior firearms and the horse, which were common along the banks of the Niger

bend to their hinterland. With the weakening of the Toucouleur Empire, the gap was filled by Samory (1830–1900) from south eastern Guinea, who became a major raider. As a child of Dyula cattle traders, his mother was captured in the course of a war, and he was sent to arrange her freedom. He therefore joined the group for over seven years and learned the use of firearms. He created his own state in 1878, and his professional army was eventually up to 35,000 strong. He traded guns for slaves with the people of Sierra Leone and Ashanti, who acquired the guns from trading with Europeans in the coastal ports. Babatu and Samory created havoc in the neighbouring countries and the Northern Territories. Samory was finally captured by the French Forces in 1898 and died two years later. Interestingly, in the Western Gonja area there were fetish villages where the horse was strictly forbidden. In 1951, the DC at Bole was asked to leave his horse outside the town (Senyon), because of the fetish. The importance of the horse was to diminish with arrival of the British, as power shifted from the ruling elites and the motor car took its place as a mode of transport. Because of trading links and invasions some northern areas had a rudimentary Islamic State structure, with Mosques and Muslim clerics. Ref. Technology and the State in Africa. See Polity and the means of Production/Ritual/ Destruction` pp.21–73 J. Goody; The N.T's of the G.C. A.E.G.Watherston. The Zabarma Conquest of North- West Ghana and Upper Volta. S.Pilaszewicz.

Gazari, followed by Babatu (Emir Babatu d'an Isa), Zabarma from Niger, came raiding. Babatu was sometimes called the ruler of the Gurunsi and Dagati and came to power in 1880, preying on the poorly defended states of the North. He died from a tarantula bite in Yendi. Fighting and raiding with Babatu were people from the Mossi, Fulani, Wangara (Dyula) and Hausa tribes. Remnants of these invading Mossi became traders, Court Interpreters or incorporated into the Mossi Horse division of the West African Force. They lived separately in a Zongo, as in Zuarungu, where Bassana even got them a separate burial ground. Both Raiders came on horseback, and used guns against the poisoned arrows of the indigenous people.

Chapter 8 - 7th Tour 1941–44

174. *Return of Fortes.* On intelligence work involving neighbouring French Territory, which was controlled by the Vichy Government, causing local British concern.

175. *Improving conditions.* Ref. The Improvement of Native Agriculture in relation to Population and Public Health. Sir A. Daniel Hall.

176. *Land for Manga Farm Centre.* The Governor at this time still passed the Ordinances, and he would have personally authorised the use of the land for Manga station. Ref. The political economy of colonialism in Ghana pp. 231–6 G.B. Kay, (Criticisms of Agricultural policy: by the West African Commission, (published 1943) 1938–9 Technical Reports, Leverhulme Trust).

177. *Tono Farm Centre.* In the original survey five centres were chosen as a pilot study. They were Tongo, Nangodi, Bongo, Sambruno and Zuarungu. This was later expanded to create three farm Centres at Zuarungu, Tono (for Navrongo) and Manga (for Bawku).

178. *Mixed farming methods* After the trip to Northern Nigeria Charles wrote Ref. Bulletin No.33 "Report on a Visit to Northern Nigeria to Study Mixed Farming". In it he describes the term "mixed farming", followed by details of its use in Northern Nigeria. It ended with a discussion of the introduction of mixed farming to the Northern Territories of the Gold Coast. The difference was Nigeria had a large agricultural staff and well- run research centres in many parts of the country. Charles aimed to introduce some of these improvements into the Northern Territories, but the lack of cash crops and therefore financial liquidity once again hindered development. Allan Kerr wrote to Fortes on 20th August 1939 about the effects of paid wages and employment in the Zuarungu District Native Authority. "As for its effect on local agriculture, a labourer now stays more permanently in a job than he did. On the other hand, Atika head mason in Zuarungu will not (in the season) leave his farm to work at 2/6d a day; nor will some lesser masons for 1/6d. They are,

incidentally, keen but not full-blown followers of Lynn.". Ref. Fortes Add. Mss 8405CUL This method of farming has been much criticised as being too slow, but many of the changes proposed cut across cultural norms, including land tenure, or could lead to the problems such as soil erosion. Agricultural change in Africa has for many decades been slow, except perhaps where commercial farming along European lines has been implemented. Charles would often say small holdings were very hard work.

179. *Sir Alan Burns* (1887—1980) from an old Colonial family served in the Leeward Islands, Nigeria, Bahamas, British Honduras and then Gold Coast 1941–7. Before coming to the Gold Coast he was an Assistant Under- Secretary of State for the Colonies, and on leaving the Gold Coast went to the UN, where he served as UK representative on the Trustee Council until 1956. The Kibi Murder case all but destroyed his career, when he was drawn into the argument with the Privy Council and Parliament over the hanging and life imprisonment debate.

180. *European children* were allowed in the South, especially on the Coast, where the climate was seen as more suitable. In 1942, the new Governor Sir Alan Burns accepted children in the North, with parents taking the responsibility. The Pienaars were children in the Boer War, interned in our concentration camps. "Varder" Pienaar worked for the railways and "Granny" Pienaar took in paying guests. They were also Sylvia's foster parents for six months, when she was just 13 months old. Florence Sibson and her daughter Gwendoline, born 12 days after Sylvia, lived together in Somerset West with the Pienaars. Isaac Sibson was then an Agricultural Officer near Accra, but later stationed in Tamale.

181. *Kapok* was also used as quilted coats for body armour instead of chain mail, before the coming of British Rule. Ref. Polity and the means of Production p.35 J Goody.

182. *John Hinds*. In a personal letter in 2003 Hinds told me this story, "In former days when lads had finished their primary education in Lawra and Wa schools they had to go to the secondary schools in Tamale. So they walked thru the bush in big groups!! I wondered what this might be like for there are

hardly any inhabitants along the way. So I trekked through, taking notes and snaps along the way. Half way along the trek I stayed at a big remote village, with a seldom used rest house a distance away under a big acacia albida tree. I hardly got any rest because of two big crows cawing hour after hour from the highest branches. I could stand it no longer. Taking my double barrel gun I shot one dead but missed the other. With that there was an immediate uproar from the vast compound. Wild ululation, drumming of every sort, trumpets, uproar. What had I done I wondered – murdered the tribal totem animal? A great crowd emerged led by the chief in all his glory. I'm for it now I thought!! But no!! The chief thanked me for slaying the crow which had been preying on the chickens for ages! But why, oh why had I missed the other equally villainous crow?" Ref. Wa and the Wala; Islam and Polity in Northwestern Ghana. I. Wilks.

183. *Censor.* All mail in the War had to pass a censor, and letters were closed again with tape showing it to have been opened by the censor. People leaving the Gold Coast also had to have exit visas. A sign of the war, and the concern over French Vichy-controlled territory around them.

184. *Marjorie's return.* Loneliness was a major factor in the rates of depression and alcoholism amongst the officers. The role of household staff was very important for their wellbeing. It was not unknown for officers to refuse advancement if it meant leaving the Gold Coast and their staff, especially if bachelors.

Chapter 9 - 8th Tour 1942–46

185. *Headquarters move to Bolgatanga.* Early Colonial DCs had sited the headquarters in Zuarungu, for pacification of the Tong Hills. Allan Kerr had long angled to get the headquarters moved, as Bolga now was on the main road, and had a very co- operative chief.

186. *Grandfather Leggott's farm* in Lincolnshire, where he spent all his childhood holidays till the untimely death of his grandfather in 1917. He especially loved the two big ploughing horses,

Blossom and Prince; it is from this time that Charles's love of nature and farming began.

187. *New Chief of Bongo,* no longer Anane Fra-fra.

188. *The Tallensi Divisional Court.* Fortes was asked by the Administration to produce a paper explaining the social context of native custom and outlining his views for the new Divisional Court in 1937. He pointed out in his guiding paper that there were no prior native institutions capable of fulfilling these functions. "Hence these proposals necessarily involve the creation of what is in effect a new institution." He suggested, "…to make use of existing native ideas, custom and institutions and out of these build something new." He pointed out how the Chiefs and Kambonaabas had taken over most of the legal functions of the society. "But underneath this system, and often in suppressed conflict with it, there is still vigorously alive the age old traditional network of loyalties and bonds based on kinship, clan organisations, religious ideas, and so forth. The vitality of this partially submerged system of traditional relationships can be gauged by the welcome which was accorded to the campaign of Col. Gibbs and the other Commissioners to re-instate family (or section) heads, in place of the Kambonaabas." Fortes goes on to define the clearly recognised clan and kin units and their close linkage. "The Tallensi method of managing all affairs of public importance is by the committee system which runs right through their social organisation down to the smallest family unit. An important decision is always taken by a committee on which every unit which it concerned is represented." From this, "A specially ingenious device, in connection with this committee system, is the distribution of key functions in such a way as to make it impossible for any one clan head – even the one recognised as the primus inter pares – to act without the co-operation of the others." He goes on to outline his ideas for consideration. Ref. Add. 8405/2/1937/8(ii)CUL.

189. *DDT* is an effective cheap chemical against mosquitoes, used extensively in the 1950s to eradicate malaria. When used as an agrochemical it affected wild life etc., so withdrawn. Diluted for indoor use, and sprayed carefully on walls, it is still a good repellent.

190. *Political Change.* In 1937, the Kunaaba and his courtiers were

sent home to Kurugu, as it was realised they were of no benefit to the Tallensi Federation, nor had real traditional authority. It was DC Nash who sent him to Beo village, as referred to in a letter to the DC Gambaga, 6th December 1911. "...3. As the last named people the Talansi (sic) are rather out of control, my chief objective in going to Gambaga was to induce the chief of Kurugu who lives in your district, to come out here and live amongst the Talansi (sic) among whom he has a good deal of influence. This he has consented to do in 6 weeks time." Provincial Commissioner E.W. Warden wrote to Nash in 1913 about problems with Kunaa's appointment, but Nash asked for more time"...let affairs take their own course." Big changes in the Zuarungu district must have come too, with the death of the Chiefs of Tongo (1941) and Zuarungu (1939), and Tengoli, the Gologdaana, in 1949. With Anane Frafra destooled in 1942 the old ways were bound to change.

Chapter 10 - 9th Tour 1946–47

191. *Lieutenant L.J. Callaghan* 1912–2005. In 1947, Callahan was appointed Parliamentary Secretary to the Minister of Transport. He was to become Chancellor of the Exchequer, Home Secretary and Foreign Secretary and eventually Prime Minister 1976–79.

192. *Tamale Airport.* Rattray landed here in a Cirrus Moth in January 1929. "Then over the Tong hills, just visible through the harmattan haze, where eight months earlier the priest had asked the god's blessing on his journey, and after an hour a white circle on the ground with 'TAMALE' written across it.; one circle over the roof of a house; flattened out; ran to within some ten yards of some huts; and came to rest. No one was expecting me; my wires had not got through, but, in what seemed after only a few seconds, old friends were all around, Hill and Harrington first, then Major and Mrs Walker Leigh, and soon the whole of Tamale." Ref. Machin p.165.

Bibliography

Adamu, Mahdi. The Hausa Factor in West African History. (ABU University Press.)

Allman, Jean & Parker, John. Tongnaab (Indiana UP).

Anafu, Moses. The Impact of Colonial Rule on Tallensi Political Institutions 1898 – 1967. (Transactions of the Ghana Historical Society)

Armitage, Capt. C.H. Notes on the Northern Territories of the Gold Coast, P. 634–9 (United Empire, The Royal Colonial Institute Journal, 1913)

Baier, Stephen. An Economic History of Central Niger.

B*arringer, Terry* How Green was our Empire (OSPA).

Binger, Capt. Louis-Gustave. Du Niger au Golfe de Guinée (Paris 1892) pp. 86–7 Salaga.

Blair, H.A. Palaver between Yamlaga and Bari Letter. 1929 (CUL Fortes Papers).

Bolton E.F. Horse Management in West Africa (Jarrold, 1931).

Broach. J. Agriculture in the Northern Territories, 1947 (Cul Fortes Papers).

Burns, Sir A. The Colonial Civil Servant. (Allen & Unwin)

Cardinall, A.W. Natives of the N.T.'s of the Gold Coast (Routledge).

Cardinall, A.W. The State of our present ethnographical knowledge of the Gold Coast People. p. 405 (Journal Africa).

Cardinall, A. W. The Gold Coast 1931 Census. (Routledge)

Clifford, Sir Hugh K.C.M.G. The Gold Coast Regiment in the East African Campaign. *1920*

Clarence-Smith, W.G. Islam and the Abolition of Slavery (Oxford UP*)*

Cooley, W.D. The Negroland of the Arabs. *(*Frank Cass 1966*)*

*Davies, Oliver. T*he invaders of Northern Ghana: what archaeologists are teaching the historians. (Universitas Achimota) Vol. IV No. 5 March 1961.

DC's Nash, Wheeler, Warden & Festing 1911–1915. (CUL Fortes Papers.)

Delavignette, Robert Louis. Freedom and Authority in French West Africa (Cass, 1968).

Destombes, Jerome. Long Term Patterns of Seasonal Hunger. Internet.

*Destombes, J*erome. Nutrition and Economic Destitution in Northern Ghana. Internet.

Dickson, K.B. A Historical Geography of Ghana (CUP).

*Dumett, Raymond & Johnson, M*arion. The End of slavery in Africa. (eds Miers & Roberts) Britain and the Suppression of Slavery in the Gold Coast Colony, Ashanti, and the Northern Territories. (Wisconsin press 1988)

Egerton, R.B. The Fall of the Asante Empire. (Free Press)

Faulkner, O.T. and Mackie, J.R., Agriculture in West Africa (CUP).
Fisher, H. J. The Horse in Central Sudan; its use. (Journal of History)
XIV 9 1973).
Forde, D. & Kaberry, P.M. West African Kingdoms in the Nineteenth
Century. (IAI Oxford)
Fortes, Meyer & S. Papers Diaries etc in (Cambridge University Library)
Mss8405
Fortes, Meyer. African Political Systems. Meyer Fortes & Edward Evans
–Pritchard (eds). (IAI Oxford Press)
Fortes, Meyer. The Dynamics of Clanship among the Tallensi (OUP).
Fortes, Meyer. The Web of Kinship amongst the Tallensi.(OUP)
Fortes, Meyer. Religion, Morality and the Person. (CUP).
Fortes, Meyer. African Cultural Values. (CUL).
Fortes, Meyer. Culture Contact as Dynamic Process. (Africa) 9, 1936.
Fortes, Meyer. Strangers. Studies in African Social Anthropology
(London. Academic Press) P229-253
Fortes. Meyer. Informants in Field Research. (CUL).
Fortes, Meyer. Divorce, Adultery and seduction (CUL).
Fortes Meyer. Marriage law amongst the Tallensi (CUL).
Fortes Meyer. Bad Arrows: an ethnographical document (CUL).
Fortes, Meyer. A Tallensi Prayer.
Fortes, Meyer. Ritual Festivities and Social Cohesion in the Hinterland of
the Gold Coast. (American Anthropologist), 1938 pp.590–604.
Fortes, Meyer. 1937. Communal Fishing and fishing Magic in the
Northern Territories of the Gold Coast (Journal Royal Anthrop Inst) 67,
pp. 131–42.
Fortes, Meyer. Towards the Judicial Process.
Fortes, Meyer. Oedipus and Job in W.A. 1959.
Fortes, Meyer. The Bonaab bagre on Tenzugu Article in Fortes paper
22/2/25.
Fortes, Meyer. A religious "Racket" in the Gold Coast. 1938 Fortes papers
(CUL).
Fortes, Meyer. Names amongst the Tallensi (German academy of Berlin)
No 26 1955.
Fortes, Meyer. Problems of Identity and Person.
Fortes, Meyer & Sonia Food in the Domestic Economy of the Tallensi.
(Africa) 9,

Garrard, T. An Asante Kudo among the Frafra. (American Museum of
History) 65.
Grischow, J.D. Shaping tradition. (Brill)
Gold Coast Coast Handbook 1937 (Ebenezer Baylis)
*Goody, J*ack. Establishing Control: Violence along the Black Volta at the
Beginning of Colonial Rule (CUP).

Goody, Jack. Technology and the State in Africa.

Goody, Jack. Literacy in Traditional Societies. Restricted Literacy in Northern Ghana.

Governors of the Gold Coast. The internet.

Gold Coast Handbook 1937

*Hall, Sir A.D*aniel. The Improvement of Native Agriculture in relation to Population and Public Health. (Oxford UP) 1936

*Hart, K*eith. The Economic Basis of Tallensi Social History in the early Twentieth Century. (JAI Press) P185-215

*Hart, K*eith. Social Anthropology of West Africa.(Ann.Rev. Anthropology) 1985 P243-72.

Hilton, T.E. Notes on the history of the Kusasi. (Historical Society of Ghana) Vol. VI 1962.

Hilton, T.E. Bawku Vol. iii No 4 1958 (Universitas.)

Hilton, T.E. Mosshi Country (Universitas) Vol. IV No. I (Dec 1959).

Hilton, T.E. Frafra resettlement and the population problem in Zuarungu Vol iii March 1959 (Universitas.)

*Hollett D*avid. The Conquest of the Niger by Land and Sea. (P.M.Heaton Publishing)

Iliasu, A.A. The establishment of British Administration in Mamprugu 1898 –1937, (Transactions of the Historical Society of Ghana) Vol. 16 1975.

Ingrams, Harold CMG, OBE. Seven across the Sahara. pp.153-160. (John Murray)

Johnson, M. The slaves of Salaga, (Journal of African History,) 27 1986 pp.341–362.

Kay, G.B. The Political Economy of Colonialism in Ghana; a collection of documents and statistics 1900-1960(CUP) 1972.

Kerr, A.F. The Frafra Constitution (CUL Fortes papers unpublished).

Kerr, A.F. Notes for Tallensi Divisional Court 1937 (CUL Fortes. unpublished).

*Kirke-Greene, A*nthony. On Crown Service. (I.B. Tauris).

Knowles, L.C.A. The Economic Development of the Overseas Empire. (Routledge).

Kuklick. H. The Imperial Bureaucrat (Hoover Colonial Studies)

Lawrence A.W. Trade Castles and Forts of West Africa   (Jonathan Cape.)

Lentz, Carola. Colonial Ethnography and Political Reform: the works of Duncan-Johnstone, R.S.Rattray, J. Eyre-Smith and J. Guinness on Northern Ghana Ufahamu (Journal of the African Activist Association) Ghana Studies 2. 1999

Levtzion, N. Asian and African studies (Israel oriental society) Vol. 2, 1966.

Lovejoy,P.E. & Hogendorn, J.S. Slow Death for Slavery. (African Studies. CUP.)

Lugard, Sir Frederick D. The Dual Mandate in British Tropical Africa (Blackwood)

Lynn, Charles W. Report on a Visit to Northern Nigeria to Study Mixed Farming. 1937 Bulletin 33. Department of Agriculture Gold Coast.

Lynn, Charles W. Agriculture in North Mamprusi, 1937 Bulletin 34 Department of Agriculture. Gold Coast.

Lynn., Charles W. Land and planning in the Northern Territories 1947

Lynn, Charles W. Agricultural Extension and Advisory Work with special references to the Colonies. (*HM* Stationary Office.)

Masefield, Geoffrey B. A short History of Agriculture in the British Colonies (Clarendon Press).

McCaskie, T.C. on R.S. Rattray and the Construction of Asante History: an appraisal. (History in Africa)10 (1983) 187–206.

.Machin, Noel "Government Anthropologist" a life of R.S. Rattray Internet publishing (Canterbury SAC Monographs.)

Mensah, A.K. The Effect of the 'Scramble' in the Gold Coast (Tamale Records).

Northcott, H.P Lt. Col. CB Report on the Northern Territories of the gold Coast. 1899. (Her Majesty's Stationary Office, by Harrison)

Nash, S.A. Visit to Gambaga to discuss Kurugu 1911. (Fortes CUL)

Nash, S.A. 1911 Report on Land Tenure. (CUL Fortes)

Ormsby- Gore, W.G. M.P. Report on North Mamprussi District for the (ultimate) information of the Honourable. 1925 DC Zuarungu. (CUL Fortes)

Page R.E. 1930. DC. Notes on Native Administration. (CUL Fortes)

Pakenham,T. The Scramble for Africa. (Abacus)

Pilaszewicz, Stanislaw. The Zabarma Conquest of North-West Ghana and Upper Volta.

Princess Marie Louise. Letters from the Gold Coast (Methuen.)

Rainsford C.C. 1926 Winkogo-Gorogo Land Dispute heard in Zuaragu before DC. (Fortes Papers)

*Rattray, Robert S*utherland. 1932 The Tribes of the Ashanti Hinterland, Vols. I and II (Oxford: Clarendon Press).

Robinson, R & Gallagher J. Africa and the Victorians (Macmillan Press)

Samory, notes off the internet.

Schnepel, B. Corporations, Personhood, and Ritual Tribal Society in Meyer Fortes. 1990.

St Clair, W. The Grand Slave Emporium. (Profile books)

Tamale; excellent undated and unsigned document papers of Fortes (CUL).

Thomas, Roger G. The 1916 Bongo "riots" and their background; aspects of Colonial Administration and African language Response in Eastern Upper Ghana. (Journal of African History) 1973

Thomas, Roger G. Forced Labour in British West Africa: the case of the Northern Territories of the Gold Coast 1906–27. (Journal of African history) 1973

Watherston, A.E.G. The Northern Territories of the Gold Coast. (Journal of the African Society) Vol. Vii (1907–8).

Wilks, Ivor. The position of Muslims in Metropolitan Ashanti in the early nineteenth Century in Islam in Tropical Africa. (IAI, Hutchinson Publishing)

Williamson, Thora. The Gold Coast Diaries Chronicles of Political Officers in West Africa 1900 -*1919*, 2000 (Radcliff Press)